10/90

£10·95

D0413617

Yet another introduction to analysis

Yet another introduction to

analysis

VICTOR BRYANT

Department of Mathematics
University of Sheffield

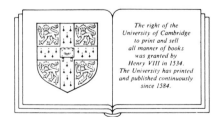

The right of the
University of Cambridge
to print and sell
all manner of books
was granted by
Henry VIII in 1534.
The University has printed
and published continuously
since 1584.

CAMBRIDGE UNIVERSITY PRESS

Cambridge

New York   Port Chester   Melbourne   Sydney

Published by the Press Syndicate of the University of Cambridge
The Pitt Building, Trumpington Street, Cambridge CB2 1RP
40 West 20th Street, New York NY 10011, USA
10 Stamford Road, Oakleigh, Melbourne 3166, Australia

© Cambridge University Press 1990

First published 1990

Printed in Great Britain by Bath Press, Bath

British Library cataloguing in publication data

Bryant, Victor
Yet another introduction to analysis.
1. Calculus
I. Title
515

Library of Congress cataloguing in publication data

Bryant, Victor.
Yet another introduction to analysis / Victor Bryant.
    p.   cm.
ISBN 0-521-38166-5.—ISBN 0-521-38835-X (pbk.)
1. Mathematical analysis.   I. Title.
QA300.B76   1990
515—dc20   89-36864   CIP

ISBN 0 521 38166 5   hard covers
ISBN 0 521 38835 x   paperback

ETA

CONTENTS

PREFACE

When I was book review editor of *The Mathematical Gazette* hardly a week went past without the arrival of yet another new book called *An introduction to analysis*. I began to think that authors and publishers were mad to invest their time and efforts in such ventures. So why on earth am I joining the ranks of those eccentrics?

Pressure upon me to write *Yet another introduction to analysis* came from 'above' and from 'below'. Following the success of my book on metric spaces (a down-to-earth but fairly rigorous approach to analysis for students in a second year at university or polytechnic) I had letters asking me to recommend a similar first-year text, and I did not know of any. But that in itself was not enough to spur me to enter this highly over-supplied market.

Mathematics education in schools has seen a revolution in recent years. Students everywhere expect the subject to be well motivated, relevant and practical. In Britain we now have the GCSE examination which has exactly those aims. My own daughter was one of the first students to take that examination and by the time students like her reach higher education not only will they have greater expectations of the subject but they will also have been less exposed to formal mathematics than their predecessors: surely our first-year university and college courses will have to take account of this?

For such students the traditional development of analysis, often rather divorced from the calculus they learnt at school, seemed highly inappropriate. Shouldn't every step in their first course in analysis arise naturally from their experience of functions and calculus? And shouldn't such a course take every opportunity to endorse and extend their basic knowledge of functions?

In this book I have tried to steer a simple and well-motivated path through the central ideas of real analysis. Each concept is introduced only after its need has become clear and after it has already been used informally. Wherever appropriate the new ideas are related to school topics and are used to extend the reader's understanding of those topics.

I assume that the reader has studied mathematics at a higher level at school (to the equivalent of A-level in England and Wales) and is familiar with the basic notation of sets and functions and elementary calculus used at that level. At each stage the reader is expected to think for her/himself and for that reason there are plenty of exercises (with solutions at the back of the book).

My sincere thanks are due to John Pym for first encouraging me to write this book and for his most helpful, constructive and open-minded advice during its preparation. My thanks also go to all the un-named colleagues, teachers, authors and contributors to the *Gazette* from whom I have gleaned so many ideas over the years.

Here then for my daughter's generation is my attempt at *yet another* introduction to analysis.

University of Sheffield   Victor Bryant

1

Firm foundations

Where do we start?

Analysis is an extension of school calculus. By building on firm foundations we intend to develop a rigorous study of the behaviour of functions. To start you off, then, try the following exercise:

$$f(x) = x^2 \quad x \in \mathbb{R}$$

Calculate $f'(1)$ and $\int_1^2 f(x)\,\mathrm{d}x$

I hope that you found that easy: $f'(1)$ (or $\mathrm{d}f/\mathrm{d}x$ at $x = 1$) is 2 and the integral is $2\frac{1}{3}$.

But what on earth do those answers mean? Aren't they something to do with gradients and area? Where did those techniques you used come from and why do they work? Try another exercise:

$$g(x) = |x| \quad x \in \mathbb{R}$$

Calculate $g'(-3)$ and $\int_{-1}^2 g(x)\,\mathrm{d}x$

You may be a little puzzled by that example as the 'modulus' function, although simple enough, is not usually included in school calculus. Never mind, consider this next example. Another function with a lot of practical uses is 'the integer part' function, where $[x]$ denotes the *integer part* of x, for example $[3.27] = 3$. Try this third exercise:

$$h(x) = [x] \quad x \in \mathbb{R}$$

Calculate $h'(2)$ and $\int_0^3 h(x)\,\mathrm{d}x$

I suspect that most of you will be giving up by now or (hopefully) beginning to think a little more about what these things mean.

Our only hope of building a solid study of analysis is by starting from firm foundations, so there's no point in plunging straight into the calculus. Let us go back to the idea of a function:

$$f(x) = x^2 \quad x \in \mathbb{R}$$
$$g(x) = |x| \quad x \in \mathbb{R}$$
$$h(x) = [x] \quad x \in \mathbb{R}$$

A function takes in a number and gives out another. For example

$$1.6 \longrightarrow \boxed{f} \longrightarrow 2.56$$

So surely to understand how functions behave we must first understand the behaviour of the numbers themselves? Our very first mathematical-looking statement above was

$$f(x) = x^2 \quad x \in \mathbb{R}$$

But what *is* $\mathbb{R}$? Our study of analysis surely *has* to start with a study of numbers.

A fresh beginning

We've now decided on a fresh start: from this point onwards we must make clear what our assumptions are and base all our deductions upon those assumptions. But how far back should I go? I could assume that you have absolutely no knowledge of numbers and begin by naming two new creations, '0' and '1', the first 'numbers'. I could then introduce an operation, ' + ', which takes in any pair of existing numbers and gives out a number as an answer. Since our only numbers so far are 0 and 1 we try those:

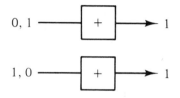

So far that is very boring indeed. But the pair of numbers fed in don't have to be different so we try:

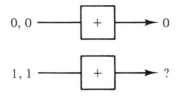

and the last example will give us a new number which we might call '2'. Then $2 + 1$ would give us a number which we might call '3', and so on. This would be a laborious way of introducing the number system. It would be rather like giving a history course which started with the formation of the first algae in the oceans: by the time you got to any interesting historical events the audience would have died from boredom. So such a basic beginning would be inappropriate here. Like the historian who, for the sake of an interesting and complete course, chooses a convenient starting-point such as the outbreak of the first world war, we are going to take a giant leap forward:

We shall assume that all the arithmetic which you met at primary school works in exactly the way you'd expect it to.

We therefore assume that there are whole numbers (or *integers*)
$$\ldots, -5, -4, -3, -2, -1, 0, 1, 2, 3, 4, 5, \ldots$$
(the set of all those is referred to as $\mathbb{Z}$) upon which the normal operations of addition, subtraction and multiplication can be applied, giving answers back in the same set. These integers have a natural order and given any two different integers, m and n say, one of them will be less than the other. If $m < n$ for example then starting at m and 'counting' m, $m + 1$, $m + 2, \ldots$ we will eventually get to n.

We also assume that any integer can be divided by any non-zero integer to give a fractional (or *rational*) number. For example,
$$2\tfrac{1}{2} \ (= 5 \div 2 = 10 \div 4 \text{ etc}), 3\tfrac{1}{7}, -22\tfrac{7}{8}, \ldots$$
and the set of all such rational numbers is denoted by $\mathbb{Q}$. These rational numbers again behave exactly as you'd expect them to from school. Any rational number m/n can be 'cancelled down' to its lowest form (for example $84/126$ would cancel down to $2/3$).

Furthermore the four arithmetic operations extend in a natural way to $\mathbb{Q}$ and behave exactly as you'd expect them to. For example,
$$3\tfrac{1}{2} + 7 = 10\tfrac{1}{2} = 7 + 3\tfrac{1}{2}$$
– the order does not matter. In general
$$a + b = b + a$$

for each of our numbers a and b (and we say that $+$ is *commutative*). Similarly

$$(2 + 5\tfrac{3}{4}) + 4\tfrac{1}{2} = 12\tfrac{1}{4} = 2 + (5\tfrac{3}{4} + 4\tfrac{1}{2})$$

$$\underbrace{\phantom{(2 + 5\tfrac{3}{4})}}_{\substack{\text{done} \\ \text{first}}} \qquad \underbrace{\phantom{(5\tfrac{3}{4} + 4\tfrac{1}{2})}}_{\substack{\text{done} \\ \text{first}}}$$

so that in general the sum $a + b + c$ makes sense without brackets, this property being called *associativity*. The same two rules hold for multiplication too. Also the two operations of multiplication and addition combine in natural ways such as with the rule

$$a(b + c) = ab + ac$$

called *distributivity*. All these rules taken together mean that brackets can be multiplied out and manipulated in the usual ways.

By making our assumptions we have taken a large leap forward and we have a viable working number system. But is it complete enough for us to be able to consider properly things like functions and differentiation etc, which is what analysis is all about? No, because even assuming all the arithmetic that we learnt at primary school there is still a huge gap in our number system.

Primary arithmetic is not enough to lead us to a full study of decimals, especially as some decimal expansions are infinite (they go on for ever) and it's not clear from primary arithmetic what that means. However, I'd now like to use decimals in an informal way to motivate where we go next. As long as we only use the decimals to motivate us then we are still keeping to our brief of basing our formal *development* on the primary arithmetic alone.

Informally then let us represent the rational numbers by their decimal expansions. For example,

$$1 = 1.0000 \ldots \text{(a 'finite' decimal, or one with 0s recurring)}$$
$$1\tfrac{1}{2} = 1.5000 \ldots \text{(again with 0s recurring)}$$
$$2\tfrac{719}{1650} = 2.43575757 \ldots \text{(with 57s recurring)}$$
$$3\tfrac{1}{7} = 3.142857142857 \ldots \text{(with 142857s recurring)}$$

All these decimals (and indeed the decimal expansions of all the numbers in $\mathbb{Q}$) recur, that is they eventually have a repeating pattern. Of course this recurring pattern may simply be of zeros, which one doesn't bother writing. But surely it's easy to make up decimals which do not have a recurring pattern? One of the most famous examples is

$$\pi = 3.141592653589793 \ldots$$

Yet our primary school construction of the arithmetic has completely missed such numbers. As a first step to incorporating them we now let $\mathbb{R}$ denote the set of all *real numbers* (which will turn out to be the set of all numbers with a decimal expansion). Whereas those numbers with recurring decimals were the rational numbers, all the rest will be called *irrational*.

We won't be sure about the existence and behaviour of the irrational numbers until we have decided which assumption to make in addition to our assumption about primary school arithmetic. But just in case you should think that the irrationals are very few-and-far-between and that they are not worth worrying too much about, try answering the following informal question:

> **Question**   Do you think that there are more rationals than irrationals, more irrationals than rationals, or about the same number of each?

Counting such huge collections is perhaps a little dodgy, so you might prefer the question in an alternative form:

> **Question**   I choose a real number between 0 and 1 at random. (For example I could have a die with ten faces and the digits 0–9 on those faces. I could then choose the decimal expansion randomly by repeatedly throwing the die.) Would the resulting number be more likely to be rational (with a constantly repeating pattern) or irrational (with no repeating pattern)?

Surely the latter is far more likely, so much so that one would hardly ever choose a rational number. So, far from being a minor gap in the real numbers, the irrational numbers will form the major part of $\mathbb{R}$.

We now pause for some exercises. In general these will form a crucial part of our development and you are urged to give them careful thought before referring to the solutions on page 221.

Exercises

1   By representing two typical rational numbers by m/n and p/q, where m, n, p and q are integers, show that the sum, difference, product and quotient of two rational numbers are all rational numbers. (Of course we shall always assume that we will only divide one number by another when the latter one is non-zero.)

Deduce that the sum and difference of a rational number r with an irrational number s gives irrational answers and that the product and quotient of a non-zero rational number r with an irrational number s gives irrational answers.

2   In one of the above examples $3\frac{1}{7} = 3.142857142857\ldots$, with a repeating string of six digits. Show why in the decimal expansion of m/n (where m and n are integers and $n > 0$) there is eventually bound to be a recurring 0 or a recurring string of less than n digits.

3   It's a surprising fact that given any positive integer n some multiple of it is of the form $99\ldots9900\ldots00$. For example given the number 74 it turns out that $135 \times 74 = 9990$. (Perhaps this is connected with the fact that

$$\tfrac{1}{74} = 0.0135135135135\ldots)$$

Prove the general result.

(Solutions on page 221).

A not-so-simple equation

Even with your basic primary school arithmetic you were soon able to move into the world of algebra and solve simple equations like

$$3 + x = 5$$
$$7x = 11$$

because in each case a straightforward arithmetic operation yields the solution. But what about the equation

$$x^2 = 2$$

How do we know that there exists an x satisfying this equation? The idea of taking square roots of 2 makes no sense if we are only allowed to assume primary school arithmetic. You could of course use your calculator and tell me that the answer is 1.414 214, but that's only an approximation since

$$(1.414214)^2 = 2.000001237796$$

In fact, since a calculator can only display a finite number of decimal places, all its answers are rational numbers. With more and more powerful calculators you might be able to obtain better and better approximations to a solution of $x^2 = 2$: e.g.

$$1.414214, 1.4142136, 1.41421357, \ldots$$

all of which give squares larger than (but closer and closer to) 2. But by this method you'll never find a number whose square *equals* 2. In fact, if you restrict attention to rational numbers (which is all you're allowed from your primary school maths) then there is no x which satisfies $x^2 = 2$, as we now see in our first 'theorem'; i.e. a result worth highlighting for future reference:

Theorem   There is no rational number x with $x^2 = 2$.

(Having stated that result we now need to prove it using only the assumptions stated earlier. Proofs of future theorems can in addition use results already established.)

Proof   Let $x = m/n$ where m and n are integers with $n > 0$. We shall take for granted the fact that the fraction m/n is written in a sensible form, with no factor in both m and n which could be cancelled down (for example we wouldn't write 38/24 when 19/12 would do).

We'll now assume that $x^2 = 2$ and we'll deduce a piece of nonsense, showing that our assumption is wrong:

$$\frac{m^2}{n^2} = \left(\frac{m}{n}\right)^2 = x^2 = 2$$

$$\therefore \ m^2 = 2n^2$$

$$\therefore \ m^2 \text{ is even}$$

But the squares of odd numbers are themselves odd. So how can the square of an integer be even? The integer itself must be even. But then its square is divisible by 4 i.e.

m **is even**

$$\therefore \ m^2 = 2n^2 \text{ is divisible by 4}$$

$$\therefore \ n^2 \text{ is even}$$

$$\therefore \ n \text{ **is even**}$$

But this means that both m and n are even, contradicting the fact that the fraction m/n was written in a sensible form.

Hence if we assume that the rational number x has $x^2 = 2$, then we get a contradiction, so there's no such rational. $\square$

Actually there's a very swish alternative way of seeing that there is no rational number which gives a sensible value of '$\sqrt{2}$' (or of '$\sqrt{q}$' unless q is a perfect square) which uses properties of prime factors. For example imagine that you thought that the rational number $x = 1.41242424\ldots$ gave $x^2 = 2$. You could then write x as a cancelled-down fraction and then express the top and bottom of that fraction as a product of primes; i.e.

$$x = 1.412424\ldots = \frac{4661}{330} = \frac{59 \times 79}{2 \times 2 \times 5 \times 5 \times 3 \times 11}$$

But then

$$2 = 1.412424\ldots^2$$
$$= \frac{59 \times 79 \times 59 \times 79}{2 \times 2 \times 5 \times 5 \times 3 \times 11 \times 2 \times 2 \times 5 \times 5 \times 3 \times 11}$$

and that's clearly impossible since *still* nothing cancels on the right-hand side. This method relies on prime factorisation and a proof based on this idea will be found in the next exercises.

So we now know that there is no rational number x with $x^2 = 2$, but how can we be sure that there *is* such an irrational number? Our primary arithmetic is no help to us when it comes to irrational numbers. In order to complete our study of $\mathbb{R}$ we need one further fundamental assumption, but it's not easy to see what form that assumption should take. We'll investigate that further in the next section.

Exercises

1   Show that $\sqrt{3}$ and $\sqrt[3]{2}$ are not rational.

2   Are you happy to accept that an eventual conclusion from primary school arithmetic is that every integer larger than 1 can be uniquely expressed as a product of prime numbers? (For example

$$1320 = 2 \times 3 \times 2 \times 5 \times 11 \times 2$$

where obviously the order of the factors is irrelevant.) If so, use this fact to show that if m and n are positive integers then it is impossible to have $m^2 = 2 \times n^2$. (Assume that m is the product of M primes, and n the product of N primes, and see what conclusion you come to.) This gives an alternative proof that $\sqrt{2}$ cannot be rational. If you're keen, show that if q is a positive integer then the only way that $\sqrt{q}$ can be rational is when q is a perfect square, in which case $\sqrt{q}$ is itself an integer.

3   (i) Suppose you are given a positive number α with $\alpha^2 > 2$. Then let

$$\beta = \frac{\alpha}{2} + \frac{1}{\alpha}$$

Show that $0 < \beta < \alpha$ and that $\beta^2 > 2$. Explain why this shows that there is no smallest positive number whose square is more than 2.

(ii) Suppose now that you are given a positive number α with $\alpha^2 < 2$. By considering $2/\alpha$ and using (i), show that there is a number β with $\beta > \alpha$ and $\beta^2 < 2$. Deduce that there is no biggest number whose square is less than 2.

(Solutions on page 223)

Piggy-in-the-middle

You will probably have met a little set theory at school: a *set* is simply a collection of objects, and for our purposes these objects will always be real numbers. In other words, all the sets which we consider are subsets of $\mathbb{R}$. Examples are

$$\mathbb{N} = \{1, 2, 3, 4, \ldots\} - \text{the set of } \textit{natural numbers}$$
$$A = \{x \in \mathbb{R}: x < 1\}$$
$$B = \{1, 1\tfrac{1}{2}, 2\}$$
$$C = \{x \in \mathbb{Q}: 0 \leqslant x \leqslant 3\}$$

The last example can be read as 'those x in $\mathbb{Q}$ – i.e. those rational x – which satisfy $0 \leqslant x \leqslant 3$'. Examples of *members* or *elements* of C are 1.5 and 2.32: we write $1.5 \in C$. It is sometimes convenient to visualise a subset of $\mathbb{R}$ informally by thinking of $\mathbb{R}$ as an infinite ruler and by marking that subset as part of the ruler. The following pictures illustrate $\mathbb{R}$ and the four sets $\mathbb{N}$, A, B and C defined above:

Some special sets which we shall encounter are called *intervals*: these are sets with the property that if two numbers are in the set then so are all the numbers between them. So the only intervals in the above examples are $\mathbb{R}$ itself and the set A. The set B, for example, is not an interval because it contains 1 and $1\tfrac{1}{2}$ but not the number $1\tfrac{3}{8}$ which is between them. Similarly C fails to be an interval because 0 and 3 are in C but, as we shall soon see, there exists an irrational number between them which is therefore not in C.

In general we write

$$[a, b] \text{ for the 'closed' interval } \{x \in \mathbb{R}: a \leqslant x \leqslant b\}$$
$$(a, b) \text{ for the 'open' interval } \{x \in \mathbb{R}: a < x < b\}$$

$(a, b]$ for the interval $\{x \in \mathbb{R}: a < x \leqslant b\}$

$(-\infty, a)$ for the interval $\{x \in \mathbb{R}: x < a\}$

$[a, \infty)$ for the interval $\{x \in \mathbb{R}: x \geqslant a\}$ etc.

Earlier we started to consider some positive real numbers whose squares are **M**ore than 2. Now let M be the set of *all* such numbers; i.e.

$$M = \{x \in \mathbb{R}: x > 0 \text{ and } x^2 > 2\}$$

Then, as we saw earlier, examples of members of M are

1.414214, 1.4142136, 1.41421357, ...

We could, in a similar way, let L be the set of all real numbers whose squares are **L**ess than 2; i.e.

$$L = \{x \in \mathbb{R}: x^2 < 2\}$$

Examples of members of L are 1.414213, 1.4142135 etc.:

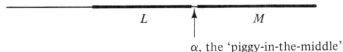

The picture of L and M shows their relative position: by sticking to positive numbers in M, L seems to be 'less than' M in some sense. And where do we expect $\sqrt{2}$ to be in relation to the two sets L and M? We would surely expect it to be 'between' them. There is no way that the existence of a number 'between' L and M can be deduced from primary arithmetic alone and so in order to proceed we need a further assumption:

> If L and M are non-empty sets with $l \leqslant m$ for each $l \in L$ and each $m \in M$, then there exists a real number α such that $\alpha \geqslant l$ for each $l \in L$ and $\alpha \leqslant m$ for each $m \in M$.

$$L \qquad\qquad\uparrow\qquad\qquad M$$

α, the 'piggy-in-the-middle'

This extra assumption is known as the *completeness axiom*: it ensures that there are 'no gaps' in $\mathbb{R}$. All analysis texts have to make an equivalent assumption somewhere and later we shall derive an equivalent form of this axiom found in many books. To see what the completeness axiom means in practice consider the following examples:

(1) $L = \{x \in \mathbb{R}: x \leqslant 2\}$ $M = \{x \in \mathbb{R}: x \geqslant 2\}$ $\alpha = 2$

In general if the non-empty sets L and M satisfy $l \leqslant m$ for each $l \in L$ and

$m \in M$ then they can have at most one number in common. If they do have a number in common then that number is the α of the axiom.

(2) $L = \{x \in \mathbb{R}: x < 1\}$ $M = \{4, 4\frac{1}{2}, 5, 5\frac{1}{2}, 6, \ldots\}$

In cases like this there are many choices for α; here any α in the interval $[1, 4]$ will do.

(3) $L = \{x \in \mathbb{R}: x^2 < 2\}$ $M = \{x \in \mathbb{R}: x > 0 \text{ and } x^2 > 2\}$

It is examples like these (which we met above) for which the completeness axiom is indispensible. In this case it ensures the existence of $\sqrt{2}$, as we now see:

> **Theorem** $\sqrt{2}$ exists.
>
> **Proof**   As above let L and M be the sets
>
> $$L = \{x \in \mathbb{R}: x^2 < 2\}   M = \{x \in \mathbb{R}: x > 0 \text{ and } x^2 > 2\}$$

Then L and M are clearly non-empty (for example $1 \in L$ and $3 \in M$) and $l \leqslant m$ for each $l \in L$ and $m \in M$. Hence by the completeness axiom there is a number α with $\alpha \geqslant l$ for each $l \in L$ and $\alpha \leqslant m$ for each $m \in M$ (and in particular α must be positive).

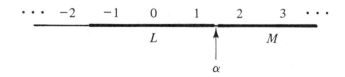

In exercise 3 on page 8 we saw that there is no smallest positive number whose square is more than 2; i.e. M has no smallest member: hence α is not in M. **Thus $\alpha^2 \leqslant 2$.**

Similarly in that exercise we saw that there is no biggest number whose square is less than 2; i.e. L has no biggest member: hence α is not in L. **Thus $\alpha^2 \geqslant 2$.**

Hence α^2 equals 2 and we have found the number α equal to $\sqrt{2}$. □

In a similar way we can prove the existence of $x^{1/n}$, the nth root of x, for any non-negative number x and any positive integer n. Hence $x^{m/n}$ ($= (x^{1/n})^m$) makes sense for the rational number m/n and, for the moment, we shall only refer to x^r when r is rational.

A common form of the completeness axiom concerns 'bounds' of sets. It will surely be immediately clear what we mean by a set being 'bounded above': in the four examples we met earlier

$$\mathbb{N} = \{1, 2, 3, 4, \ldots\}$$
$$A = \{x \in \mathbb{R} : x < 1\}$$
$$B = \{1, 1\tfrac{1}{2}, 2\}$$
$$C = \{x \in \mathbb{Q} : 0 \leqslant x \leqslant 3\}$$

the sets A, B and C are bounded above but $\mathbb{N}$ does not seem to be. (We shall verify that fact in the next exercises. Don't ever be tempted to think that ∞ is a number – what on earth would its decimal expansion be?) In general a set E is *bounded above* if there exists a number u (which will be called an *upper bound*) such that $e \leqslant u$ for all e in E. The set A given here is bounded above, and examples of its upper bounds are 4, 270 and $1\tfrac{1}{2}$ (there are lots of them! – we didn't say that u was unique) although perhaps the most natural choice of upper bound would be the number 1, being the smallest of all the possible upper bounds of A. The smallest upper bound of a set E is sometimes called the *supremum* of E, and is denoted by $\sup E$ (it's called *the* supremum because if there is one it *is* unique). In the four examples above the sets A, B and C were non-empty and bounded above. Their suprema (= plural of supremum) are given by

$$\sup A = 1 \qquad \sup B = 2 \qquad \sup C = 3$$

Can we now be sure, with the aid of our new assumption, that non-empty sets which are bounded above always have a smallest upper bound (or supremum)? We now prove that result, often used in text books in place of our completeness axiom:

Theorem   If L is a non-empty set which is bounded above then L has a supremum.

Proof   Let L be a non-empty set which is bounded above and let M be the set of upper bounds of L. Then as L has at least one upper bound it follows that the set M is non-empty. In addition it is clear that $l \leqslant m$ for each $l \in L$ and each $m \in M$.

We can now apply the completeness axiom to L and M to deduce that there exists a number α with $\alpha \geqslant l$ for each $l \in L$ (so that α is an *upper bound* of L) and $\alpha \leqslant m$ for each $m \in M$ (so that α is the *smallest* of all the upper bounds).

Hence α is the least upper bound (or supremum) of L.   □

Of course there is nothing special about the top end of a set: we can define what we mean by a set being 'bounded below' and look for its largest lower bound and we can talk about a set being *bounded*, which means that it is both bounded above and bounded below.

Exercises

1   Given any positive number u let $[u]$ denote its 'integer part'. Show that $[u] + 1$ is a positive integer which is larger than u. Deduce that the set $\mathbb{N}$ of positive integers has no upper bound.

2   (i) Which of these sets are bounded above? What are their suprema?

$$A = \{\text{the prime numbers}\}$$

$$B = \{1/p: p \text{ is a prime number}\}$$

$$C = \{\tfrac{1}{2}, \tfrac{2}{3}, \tfrac{3}{4}, \tfrac{4}{5}, \tfrac{5}{6}, \tfrac{6}{7}, \tfrac{7}{8}, \tfrac{8}{9}, \tfrac{9}{10}, \ldots\}$$

$$D = \{1, \tfrac{2}{3}, \tfrac{3}{5}, \tfrac{4}{7}, \tfrac{5}{9}, \tfrac{6}{11}, \tfrac{7}{13}, \tfrac{8}{15}, \ldots\}$$

(ii) Show that if A and B are any non-empty sets with $a + b \leqslant \alpha$ for each $a \in A$ and $b \in B$ then A and B are bounded above and

$$\sup A + \sup B \leqslant \alpha$$

3   (i) Show that a non-empty set M which is bounded below has a biggest lower bound (called the *infimum* and denoted by $\inf M$).

(ii) Which of the sets in exercise 2(i) are bounded below? What are their infima?

(iii) Now let L and M be any non-empty sets with $l \leqslant m$ for each $l \in L$ and $m \in M$ and such that there is a *unique* number α 'between' L and M (i.e. with $l \leqslant \alpha \leqslant m$ for each $l \in L$ and $m \in M$). Show that

$$\alpha = \sup L = \inf M$$

(iv) Let A be a non-empty set which is bounded below and let B be the set $\{-a: a \in A\}$. Show that B is bounded above and that

$$\sup B = -\inf A$$

4   (i) Let E be a set and consider the following two statements about E:

(1) there exists a number b with $-b \leqslant e \leqslant b$ for all $e \in E$;

(2) E is bounded (i.e. bounded above and bounded below).

Show that if (1) is true then so is (2), and that if (2) is true then so is (1). (Mathematicians say that (1) happens **if and only if** (2) does.)

(ii) Prove that the union of two bounded sets is bounded. (Remember that the union $E \cup E'$ of the sets E and E' consists of all the elements which were in E or in E' or both.)

5   Let x and y be different real numbers and let $0 < \beta < 1$. Show that the number $x + \beta(y - x)$ is between x and y. By choosing appropriate values of β show that between any two different rational numbers there are both rational numbers and irrational numbers, and that between a rational and an irrational number there is an irrational number.

(We shall deduce in the next exercises that between any two different numbers there are both rationals and irrationals.)

6   One of the following school exercises is wrong: which one, and why?

(i) The attendance at a football match is 23 000 to the nearest thousand. What is the largest number of people that could have been at the match?

(ii) An angle is measured to the nearest degree and found to be 48°. What is the largest possible value of the angle?

(Solutions on page 224)

Some natural consequences

Sometimes a set may have a biggest or a smallest member, known as its *maximum* or *minimum*: note carefully that the maximum or minimum of a set has to be a member of the set. For example the set

$$\{x \in \mathbb{R}: 0 < x \leqslant 3\}$$

has a maximum of 3 but no minimum member (if you give me *any* number in the set then I can find a smaller one by halving your number), whereas the set

$$\{2, 4, 6, 8, 10, \ldots\}$$

has a minimum but no maximum. In exercise 6 above the set

$$\{n \in \mathbb{N}: n \text{ to the nearest thousand is 23 000}\}$$

has a maximum member 23 499 but the set

$$\{x \in \mathbb{R}: x \text{ to the nearest whole number is 48}\}$$

had no maximum. It seems that bounded sets of integers are more likely to have maximum and minimum members than arbitrary sets of numbers, as we now establish.

> ***Theorem*** Any non-empty set of integers which is bounded above has a maximum member. Similarly, any non-empty set of integers which is bounded below has a minimum member. In particular any non-empty subset of the set $\mathbb{N}$ of positive integers has a minimum member.

Proof   Let E be any non-empty set of integers which is bounded above. Then E contains some integer e and also the set E has some upper bound u. Let e' be any integer larger than u. Then working through the decreasing list of integers $e', e' - 1, e' - 2, \ldots, e + 1, e$, we come to a first one which is in E. This is clearly the maximum member of E. Similarly if E' is any non-empty set of integers which is bounded below then there exists an integer e in E' and another integer e' which is a lower bound of E'. Then working through the list of integers $e', e' + 1, e' + 2, \ldots, e - 1, e$, it is clear that the first one which is in E' is the minimum member of E'.

In particular any non-empty subset of $\mathbb{N}$ is bounded below (by 1 for example) and it follows that such a set will have a minimum member. $\square$

The last part of that result is the cornerstone of 'the principle of mathematical induction', but before deducing that principle as a 'corollary' (or consequence) of the theorem we illustrate it by means of an example.

Example   Let the numbers $x_1, x_2, x_3, \ldots$ be defined by

$x_1 = 2$ and $x_n = \sqrt{(6 + x_{n-1})}$ for $n > 1$.

Show that each of the numbers satisfies $2 \leqslant x_n < 3$.

Solution   Consider the required property '$2 \leqslant x_n < 3$' as a property **of n**. So $n = 1$ has the property since $2 \leqslant x_1 < 3$.

Hence

$$2 \leqslant x_1 < 3$$

But then

$$2 \leqslant \underbrace{\sqrt{(6 + 2)} \leqslant \sqrt{(6 + x_1)}}_{= x_2} < \sqrt{(6 + 3)} = 3$$

i.e.

$$2 \leqslant x_2 < 3$$

and so $n = 2$ has the property. But then

$$2 \leqslant \underbrace{\sqrt{(6 + 2)} \leqslant \sqrt{(6 + x_2)}}_{= x_3} < \sqrt{(6 + 3)} = 3$$

i.e.

$$2 \leqslant x_3 < 3$$

and so $n = 3$ has the property. But then

$$\vdots$$

i.e.

$$2 \leqslant x_{k-1} < 3$$

and so $n = k - 1$ has the property. But then

$$2 \leqslant \underbrace{\sqrt{(6 + 2)} \leqslant \sqrt{(6 + x_{k-1})}} < \sqrt{(6 + 3)} = 3$$

$$= x_k$$

i.e.

$$2 \leqslant x_k < 3$$

$$\vdots$$

and so the result holds for all positive integers.   □

Once we have established the general 'principle of induction' the examples will be rather easier.

Corollary (*The principle of mathematical induction*) Let P be a property which may or may not hold for any positive integer n, and assume that:

(I) the number 1 has the property P;

(II) if any positive integer $k - 1$ has the property then so does the integer k.

Then it follows that the property P holds for all positive integers.

Proof Let F be the set of positive integers for which the property fails (we hope to show that F is the empty set).

By (I) the property holds for the number 1 and so that number is not in F. We can now deduce from the given conditions (I) and (II) the following fact about F:

If k is in F then $k - 1$ is also in F

(For if k is in F then $k > 1$ and by (II) we cannot have the property P holding for $k - 1$ but not for k.) Hence the set F has no minimum member. By the theorem the only subset of $\mathbb{N}$ which does not have a minimum member is the empty set. It follows that F is the empty set and that the property P holds for all positive integers.   □

Example Show that for each positive integer n and for each real number $x \neq 1$

$$1 + x + x^2 + \cdots + x^{n-1} = \frac{1 - x^n}{1 - x}$$

Solution   Although there is an alternative method we shall use induction to establish the sum of this 'geometric progression' (or 'geometric series').

(I) In the case $n = 1$ the result is trivial, both sides of the required equation being 1.

(II) If the result is known for some positive integer $k - 1$ then

$$1 + x + x^2 + \cdots + x^{k-2} = \frac{1 - x^{k-1}}{1 - x}$$

and so

$$1 + x + x^2 + \cdots + x^{k-1}$$
$$= (1 + x + x^2 + \cdots + x^{k-2}) + x^{k-1}$$
$$= \frac{1 - x^{k-1}}{1 - x} + x^{k-1} = \frac{1 - x^{k-1} + x^{k-1} - x^k}{1 - x} = \frac{1 - x^k}{1 - x}$$

which is the required result in the case when $n = k$. Hence if the positive integer $k - 1$ works then so does k.

By the principle of induction established above the result therefore holds for every positive integer n.   □

Example   Show that any finite set (i.e. one which contains a finite number of members) is bounded.

Solution   We aim to show that the set

$$\{x_1, x_2, \ldots, x_n\}$$

is bounded. You might think that this is obvious because you can put the n numbers in order and then the lowest forms a lower bound of the set and the highest forms an upper bound of the set. But in case you feel that such an argument involving 'n' items is a little dodgy we'll prove this result by induction.

(I) In the case $n = 1$ the set is $\{x_1\}$ which is clearly bounded below (by x_1) and bounded above (by x_1 too!).

(II) So now assume that the result is known for the positive integer $k - 1$; i.e. any set of $k - 1$ numbers is bounded. Consider the case when $n = k$:

$$\{x_1, x_2, \ldots, x_k\} = \underbrace{\{x_1, x_2, \ldots, x_{k-1}\}}_{\substack{\text{bounded (by our assumption} \\ \text{concerning the } k-1 \text{ case)}}} \cup \underbrace{\{x_k\}}_{\substack{\text{bounded} \\ \text{(by (I))}}}$$

which is bounded since (as we saw in exercise 4 above) the union of two bounded sets is bounded.

That completes the proof by induction and shows that a finite set $\{x_1, x_2, \ldots, x_n\}$ is bounded. $\square$

Exercises

1   Which of the following sets has a maximum member and which has a minimum member?

$\varnothing$ = the empty set

$A = \{x \in \mathbb{Q}: 1 \leqslant x \leqslant \sqrt{2}\}$

$B = \{1/p: p \text{ is a prime})$

$C = \{1, 1 + \frac{1}{2}, 1 + \frac{1}{2} + \frac{1}{4}, 1 + \frac{1}{2} + \frac{1}{4} + \frac{1}{8}, 1 + \frac{1}{2} + \frac{1}{4} + \frac{1}{8} + \frac{1}{16}, \ldots\}$

$D = \{1, 1 - \frac{1}{3}, 1 - \frac{1}{3} - \frac{1}{9}, 1 - \frac{1}{3} - \frac{1}{9} - \frac{1}{27}, 1 - \frac{1}{3} - \frac{1}{9} - \frac{1}{27} - \frac{1}{81}, \ldots\}$

2   Show that if a set has a maximum member then it has a supremum which is contained in the set. Conversely show that if the set has a supremum which is contained in it then the set has a maximum member. (So that a set has a maximum member **if and only if** it has a supremum contained in the set.)

State and prove a result connecting the infimum and the minimum of a set.

3   Let x and y be real numbers with $y > x + 1$. By considering the largest member of the set

$$\{n \in \mathbb{Z}: n < y\}$$

show that there exists an integer between x and y.

4   We shall now try to establish that between any two different numbers there is a rational number. To try to see first an informal verification of this fact imagine for a moment that the two numbers are more than $1/101$ apart. Then surely at least one of the fractions

$$\ldots \frac{-3}{101}, \frac{-2}{101}, \frac{-1}{101}, 0, \frac{1}{101}, \frac{2}{101}, \frac{3}{101}, \ldots$$

illustrated below would lie between the two numbers?

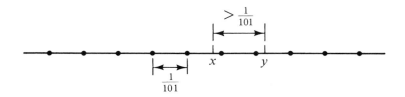

Formally suppose that x and y are two real numbers with $x < y$. Let N

be an integer chosen with $N > 1/(y - x)$. (How do we know that such an integer exists?) Use the previous exercise to show that there exists an integer M with $Nx < M < Ny$ and deduce that there is a rational number between x and y.

Deduce that between any two different numbers there is a rational and an irrational number.

5   Prove by induction that for each positive integer n

(i) $\dfrac{1}{1 \times 2} + \dfrac{1}{2 \times 3} + \dfrac{1}{3 \times 4} + \cdots + \dfrac{1}{n(n + 1)} = 1 - \dfrac{1}{n + 1}$

(ii) $1^2 + 2^2 + 3^2 + \cdots + n^2 = \frac{1}{6}n(n + 1)(2n + 1)$

6   Prove by induction that for each positive integer n

(i) $(1 + x)^n \geqslant 1 + nx$

where x is any number with $x \geqslant -1$. (Think carefully about where in your solution you need the fact that $x \geqslant -1$.)

This result is known as *Bernoulli's inequality* and is credited to Jacob Bernoulli, one of a great Swiss mathematical family of the seventeenth and eighteenth centuries.

(ii) $(1 + x)^n = \dbinom{n}{0} + \dbinom{n}{1}x + \dbinom{n}{2}x^2 + \cdots + \dbinom{n}{r}x^r + \cdots + \dbinom{n}{n}x^n$

where x is any number and $\dbinom{n}{r}$ denotes the 'binomial coefficient' and equals $n!/(r!(n - r)!)$. (You will need the fact that

$$\binom{k - 1}{r - 1} + \binom{k - 1}{r} = \binom{k}{r}$$

which you can easily verify for yourselves.)

This result is known as the *binomial theorem*.

7   Prove by induction that for each positive integer n

(i) $1 + \dfrac{1}{2} + \dfrac{1}{3} + \dfrac{1}{4} + \dfrac{1}{5} + \cdots + \dfrac{1}{2^{n-1}} \geqslant \dfrac{1}{2}(n + 1)$

(ii) $1 + \dfrac{1}{2^r} + \dfrac{1}{3^r} + \dfrac{1}{4^r} + \dfrac{1}{5^r} + \cdots + \dfrac{1}{(2^n - 1)^r} \leqslant \dfrac{1 - (\frac{1}{2})^{(r-1)n}}{1 - (\frac{1}{2})^{r-1}} \quad (r \neq 1)$

(Solutions on page 227)

Some loose ends

Our final topics in this chapter concern two numbers about which you may be a little vague (if you're anything like I was when I started studying analysis) and yet which keep popping up in our studies: they are the numbers e and π. I used to wonder why 'logs to the base e' are called the *natural* logs when 10 seemed a much more natural base than e (≈ 2.71828). And why did we switch from the nice neat $180°$ in a straight line to the peculiar π (≈ 3.14159) 'radians'? These are questions which we shall answer in due course for it takes a fair bit of analysis before we can see these two numbers at work. But having struggled to set up the real numbers and to pin-point $\sqrt{2}$ it is fitting to end the chapter by establishing the existence of these two numbers.

Theorem   Let E be the set of numbers

$$\left\{1, 1 + \frac{1}{1!}, 1 + \frac{1}{1!} + \frac{1}{2!}, 1 + \frac{1}{1!} + \frac{1}{2!} + \frac{1}{3!}, 1 + \frac{1}{1!} + \frac{1}{2!} + \frac{1}{3!} + \frac{1}{4!}, \ldots\right\}$$

Then E is bounded above. Its least upper bound is called e and it is irrational.

Proof   Note that

$$3! = 3 \times 2 > 2^2, \qquad 4! = 4 \times 3 \times 2 > 2^3, \qquad 5! = 5 \times 4 \times 3 \times 2 > 2^4 \text{ etc}$$

(and if you were very fussy you could establish the general result by using induction). Hence, by using the sum of a geometric progression (established in an earlier example) in the case $x = \frac{1}{2}$, we see that for $n > 2$

$$1 + \frac{1}{1!} + \frac{1}{2!} + \frac{1}{3!} + \cdots + \frac{1}{n!} < 1 + 1 + \frac{1}{2} + \frac{1}{4} + \cdots + \frac{1}{2^{n-1}}$$

$$= 1 + \frac{1 - (\frac{1}{2})^n}{1 - \frac{1}{2}} = 3 - \frac{1}{2^{n-1}} < 3$$

Therefore every member of E is less than 3 and the set E is bounded above. It follows that E has a least upper bound, which we shall denote by e.

Before proceeding to prove that e is irrational note that for any integer $N > 2$

$$\frac{1}{(N+1)!} + \frac{1}{(N+2)!} + \frac{1}{(N+3)!} + \cdots + \frac{1}{(N+k)!}$$

$$= \frac{1}{(N+1)!}\left(1 + \frac{1}{(N+2)} + \frac{1}{(N+2)(N+3)} + \cdots\right.$$

$$\left. + \frac{1}{(N+2)(N+3)\ldots(N+k)}\right)$$

$$< \frac{1}{(N+1)!}\left(1 + \frac{1}{(N+2)} + \frac{1}{(N+2)^2} + \cdots + \frac{1}{(N+2)^{k-1}}\right)$$

$$< \frac{1}{(N+1)!}\left(1 + \frac{1}{2} + \frac{1}{4} + \cdots + \frac{1}{2^{k-1}}\right)$$

$$< \frac{2}{(N+1)!} = \frac{2}{N+1}\cdot\frac{1}{N!} \leqslant \frac{1}{2N!}$$

(the upper bound of 2 for the geometric progression following easily from the sum of a geometric progression in the earlier example in the case $x = \frac{1}{2}$).

We shall now show that e is irrational by assuming that it is rational and deducing a contradiction.

Assume then that e is the rational m/n where m and n are positive integers. Then by multiplying the top and bottom of this fraction by appropriate factors we can deduce that

$$\text{e} = \frac{M}{N!} \text{ for some integers } M \text{ and } N \text{ with } N > 2$$

For example if e = 3 then e = 18/3! and if e = 14/5 then e = 336/5! Now

$$1 + \frac{1}{1!} + \frac{1}{2!} + \frac{1}{3!} + \cdots + \frac{1}{N!} = \frac{P}{N!} \text{ say}$$

is a member of the set E. Also from the inequalities established above we can deduce that any larger member of the set E satisfies

$$1 + \frac{1}{1!} + \frac{1}{2!} + \frac{1}{3!} + \cdots + \frac{1}{(N+k)!}$$

$$= \underbrace{1 + \frac{1}{1!} + \frac{1}{2!} + \frac{1}{3!} + \cdots + \frac{1}{N!}}_{=\frac{P}{N!}}$$

$$+ \underbrace{\frac{1}{(N+1)!} + \frac{1}{(N+2)!} + \cdots + \frac{1}{(N+k)!}}_{<\frac{1}{2N!}} < \frac{P + \frac{1}{2}}{N!}$$

Let us digest those facts: $P/N!$ is a member of the set E and although there are bigger members than this none exceeds $(P + \frac{1}{2})/N!$. Let us illustrate the numbers of the form $1/N!, 2N!, 3/N!, \ldots$ together with members of the set E and the number e:

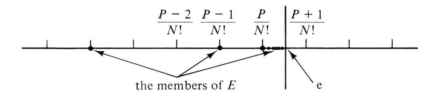

the members of E

Thus $e = M/N!$, the least upper bound of the set E satisfies

$$\frac{P}{N!} < \frac{M}{N!} < \frac{P+1}{N!}$$

which is impossible since M is an integer. This contradiction shows that e is irrational as claimed.   □

As mentioned earlier the number e will occur many times in our studies. The numbers in the set E defined above increase towards the supremum e, getting closer and closer to it. In exercise 1 below you are asked to use this fact to calculate an approximate value of e. The existence of π is derived in a similar way in the exercises.

Surely after all this work building the real numbers on firm foundations, we shall be ready to study functions in the next chapter?

Exercises

1   Use your calculator (or computer) to work out the approximate value of the first few members of the set E in the above theorem. Hence obtain an approximation for e.

2   Let P be the set of numbers

$$\{\sqrt{6}, \sqrt{(6(1 + \tfrac{1}{4}))}, \sqrt{(6(1 + \tfrac{1}{4} + \tfrac{1}{9}))}, \sqrt{(6(1 + \tfrac{1}{4} + \tfrac{1}{9} + \tfrac{1}{16}))},$$
$$\sqrt{(6(1 + \tfrac{1}{4} + \tfrac{1}{9} + \tfrac{1}{16} + \tfrac{1}{25}))}, \ldots\}$$

Use the fact that $1/m^2 < 1/m(m - 1)$ together with exercise 5(i) on page 19 to show that the number $\sqrt{12}$ is an upper bound of P. The members of P are increasing towards its supremum: use your calculator or computer to work out some of the members of P and hence obtain an estimate of $\sup P$. (In fact the supremum is π, which arises in other much more natural ways as we shall see in later chapters. And we shall prove in the very last exercise of the book that π is irrational.)

3   In the two previous exercises we have used our calculators to obtain approximations to the supremum of a set, but only after we had proved mathematically that the supremum existed (by showing that each

set was bounded above). It is dangerous to rely on intuition based on a few calculations when trying to show that a set is bounded above, and the point of this next exercise is to endorse that fact.

Let E be the set

$$\{1, 1 + \tfrac{1}{2}, 1 + \tfrac{1}{2} + \tfrac{1}{3}, 1 + \tfrac{1}{2} + \tfrac{1}{3} + \tfrac{1}{4}, 1 + \tfrac{1}{2} + \tfrac{1}{3} + \tfrac{1}{4} + \tfrac{1}{5}, \ldots\}$$

Calculate a few terms of E. Do you think that the set is bounded above?

In exercise 7(i) on page 19 we showed that

$$1 + \frac{1}{2} + \frac{1}{3} + \frac{1}{4} + \frac{1}{5} + \cdots + \frac{1}{2^{n-1}} \geq \frac{1}{2}(n+1)$$

Use this fact to show that E is not bounded above.

Use part (ii) of that same exercise to show, however, that for $r > 1$ the set

$$\left\{1, 1 + \frac{1}{2^r}, 1 + \frac{1}{2^r} + \frac{1}{3^r}, 1 + \frac{1}{2^r} + \frac{1}{3^r} + \frac{1}{4^r}, 1 + \frac{1}{2^r} + \frac{1}{3^r} + \frac{1}{4^r} + \frac{1}{5^r}, \ldots\right\}$$

is bounded above.

(Solutions on page 231)

2

Gradually getting there

Calculus at last?

We have now put the study of the real numbers onto a firm footing and we are ready to consider the role of a function and its effect on numbers. For example let the function f be given by

$$f(x) = x^2 \quad x \in \mathbb{R}$$

As you will already know, this means that f will accept any x in $\mathbb{R}$ (called the *domain* of f) and give out the answer x^2. You will also be familiar with the idea of the 'graph' of a function. Take the usual coordinate (x, y)-plane and, for each x in f's domain, mark the point $(x, f(x))$ in that plane. The resulting collection of all such points is the *graph* of f.

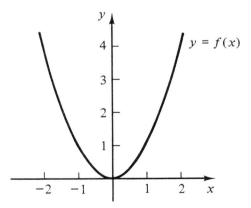

We started the first chapter with that particular function f and recalled by A-level techniques that

$$f'(1) = 2 \quad \text{and} \quad \int_1^2 f(x)\, dx = 2\tfrac{1}{3}$$

But we then had to postpone further discussion until we had looked at the structure of $\mathbb{R}$. But now we can begin to consider the meaning of those answers. How would you explain in down-to-earth terms what '$f'(1) = 2$' means? You might say that it is the gradient when $x = 1$ on the graph drawn above; i.e. at the point P with coordinates $(1, 1)$. But even the idea of the gradient of a *curve* is not trivial: are we expected to draw free-hand a tangent to the curve at P?

Let us now consider an accurate and analytical way of finding the gradient of f at $P(1, 1)$. The only gradients which make immediate sense are those of straight lines. So to get a first approximation to the value of the gradient **at** P we'll instead calculate the gradient of the straight line from P to some other nearby point Q on the graph of f, the line PQ being called a *chord* of the graph. For example let us consider the point Q for which $x = 1.8$, namely the point $Q(1.8, 3.24)$.

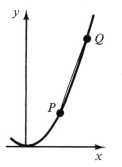

In this figure P is the point $(1, 1)$ and Q is the point $(1.8, 3.24)$ and so the gradient of the chord PQ is

$$\frac{\text{change in } y}{\text{change in } x} = \frac{3.24 - 1}{1.8 - 1} = 2.8$$

As you can see from the picture the gradient of PQ is approximately that of the graph at P. We can improve this approximation by endless choices for Q closer and closer to P, and the next three pictures show some possibilities for the first few choices of Q.

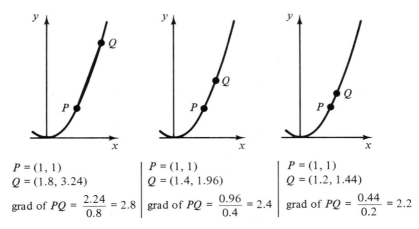

$P = (1, 1)$	$P = (1, 1)$	$P = (1, 1)$
$Q = (1.8, 3.24)$	$Q = (1.4, 1.96)$	$Q = (1.2, 1.44)$
grad of $PQ = \dfrac{2.24}{0.8} = 2.8$	grad of $PQ = \dfrac{0.96}{0.4} = 2.4$	grad of $PQ = \dfrac{0.44}{0.2} = 2.2$

You might already be thinking that this shows that the gradient at P is 2, but although you can choose Q closer and closer to P, Q must never *equal* P for then the idea of the gradient of PQ would make no sense. So the best we can do is to move Q closer and closer to P (taking, for example, x equal to 1.8, 1.4, 1.2, 1.1, 1.05, ...) and to get a list (or *sequence*) of values

$$2.8, \ 2.4, \ 2.2, \ 2.1, \ 2.05, \ \ldots$$

which are 'tending towards' the value of the gradient at P. It seems clear that, in some sense, this sequence is tending towards 2 and so perhaps that is the value we choose as $f'(1)$, the gradient of f **at** P.

Now let us look at

$$\int_1^2 f(x)\,\mathrm{d}x = 2\tfrac{1}{3}$$

Doesn't that $2\tfrac{1}{3}$ represent the area under the graph of f from $x = 1$ to $x = 2$?

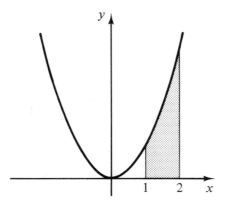

How are we meant to calculate that area from basic principles? It's hardly practicable to draw the graph on squared paper and count the squares in the required area, but we can imitate that process exactly and analytically.

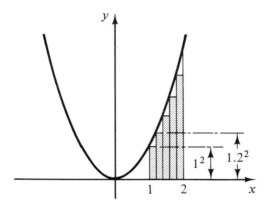

We first divide the area into vertical strips or columns, the case of five strips being illustrated above. Then we form the biggest rectangles possible which will fit under the curve, as shown. The total area of these five rectangles is

$$(0.2 \times 1^2) + (0.2 \times 1.2^2) + (0.2 \times 1.4^2) + (0.2 \times 1.6^2) + (0.2 \times 1.8^2) = 2.04$$

and that is a first approximation to the required area under the curve. To improve that approximation we could divide the area into more and more strips: a first few possible examples are shown below, doubling the number of strips each time.

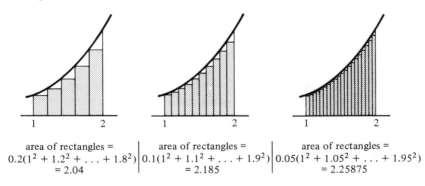

area of rectangles =
$0.2(1^2 + 1.2^2 + \ldots + 1.8^2)$
$= 2.04$

area of rectangles =
$0.1(1^2 + 1.1^2 + \ldots + 1.9^2)$
$= 2.185$

area of rectangles =
$0.05(1^2 + 1.05^2 + \ldots + 1.95^2)$
$= 2.25875$

So again we don't yet have a concrete answer for the area but a sequence of values which will tend to the required answer, namely

$$2.04, \ 2.185, \ 2.25875, \ 2.29594, \ 2.31461, \ldots$$

Even in this simple example the precise value to which that sequence is tending is not immediately clear.

To summarise, then, for both the above problems in differentiation and integration we found a fairly precise analytical process which led not to a single answer but a list or *sequence* of answers tending in the right direction. So I'm sorry to have to postpone yet again our study of calculus, but from what we have seen it is clear that we need to study sequences as our next step towards a full study of functions.

Exercises

1   The figures show the graph of the function g given by

$$g(x) = \frac{1}{x} \quad x > 0$$

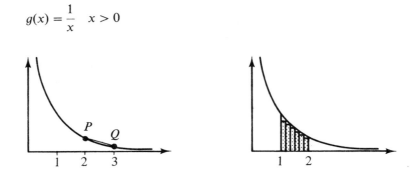

(i) The left-hand figure shows the chord from $P\ (2, \frac{1}{2})$ to $Q\ (3, \frac{1}{3})$ on the curve. Calculate the gradient of PQ. Now keep P fixed and choose other values of Q on the curve but closer to P (for example you could take, in turn, those Q with x-coordinates 2.5, 2.2, 2.1, and more if you can be bothered). Calculate the gradient of the chord PQ in each case and hence obtain the first few terms of a sequence which is tending towards 'the gradient of g at P'.

(ii) The right-hand figure shows the area under the curve between $x = 1$ and $x = 2$ divided into 6 strips of equal width. Use your calculator to work out the total area of the 6 rectangles illustrated. Repeat the process for the area divided into more and more strips (for example you could try 8, 10 and 12 strips, and more if you could be bothered). Hence obtain the first few terms of a sequence which is tending towards 'the area under the graph of g from $x = 1$ to $x = 2$'.

2   In chapter 1 we calculated the number e as approximately 2.718 282. Let g be the same function as in exercise 1: we wish to obtain an approximate value of the area under the graph of g from $x = 1$ to $x = $ e.

Divide that area into 10 (or 1000?) strips and form rectangles as in the previous exercise, and use your calculator (or ideally a computer) to calculate the approximate total area of those rectangles. Repeat the procedure for larger and larger numbers of strips. Hence obtain some terms of a sequence which is tending towards 'the area under the graph of $g(x) = 1/x$ from $x = 1$ to $x = e$'.

3   We proved in chapter 1 that $\sqrt{2}$ exists although we didn't specifically calculate it. We also saw, in exercise 3 on page 8, that if α is a positive number with $\alpha^2 > 2$ then the number $\beta = \alpha/2 + 1/\alpha$ is also positive with $\beta^2 > 2$ and with β^2 closer to 2. Choose any number x_1 which is clearly over $\sqrt{2}$ (e.g. $x = 10$!) and use your calculator to work out the new number

$$x_2 = \frac{1}{x_1} + \frac{x_1}{2}$$

By the above comments this will be closer to $\sqrt{2}$. Then calculate

$$x_3 = \frac{1}{x_2} + \frac{x_2}{2}, \quad x_4 = \frac{1}{x_3} + \frac{x_3}{2}, \quad x_5 = \frac{1}{x_4} + \frac{x_4}{2}, \ldots$$

and in this way obtain the first few terms of a sequence which seems to tend towards $\sqrt{2}$.

(Solutions on page 232)

Sequences

We concluded at the end of the previous section that we must study sequences before we can study functions fully. In fact a sequence is a special sort of function. Consider the functions

$$f(x) = x^2 \quad x \in \mathbb{R}$$

$$g(x) = \sqrt{x} \quad x \geq 0$$

$$h(x) = \frac{x}{x + 1} \quad x \in \mathbb{N}$$

You will be familiar with examples like the first two, where the domains are huge chunks of $\mathbb{R}$, but perhaps a little more surprised by h's domain being the set of natural numbers $\{1, 2, 3, \ldots\}$. Let's illustrate the graphs of these functions:

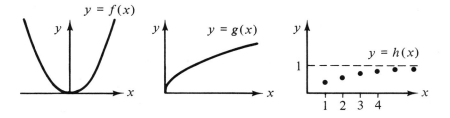

In the third case h can be stipulated as a list of numbers $h(1)$, $h(2)$, $h(3)$, ..., perhaps stating what the nth one will be in case there is any doubt; i.e.

$$\frac{1}{2}, \frac{2}{3}, \frac{3}{4}, \frac{4,}{5}, \frac{5}{6} \quad \cdots \quad \frac{n}{n+1}, \quad \cdots$$

We are then back to the simple idea of a sequence being a list of numbers and although a sequence can be poshly thought of as a 'function with domain $\mathbb{N}$' the down-to-earth idea of a list suits us very well. In general our sequences will always be of real numbers and we shall write a sequence in the form

$$x_1, x_2, x_3, \ldots, x_n, \ldots$$

(Many authors abbreviate this to (x_n) or $\{x_n\}$ but until you are very familiar with the concept of a sequence the fuller form $x_1, x_2, x_3, \ldots$ is more dynamic and easy to comprehend.)

Sometimes the graph of a sequence (like the graph of h above) does help us to see the behaviour of a particular sequence. For example that graph of

$$\frac{1}{2}, \frac{2}{3}, \frac{3}{4}, \frac{4}{5}, \frac{5}{6} \quad \cdots \quad \frac{n}{n+1}, \quad \cdots$$

does seem to imply that the sequence is 'increasing and getting closer and closer to 1'.

Some examples of sequences are

(i) $1, 2, 3, 4, \ldots n, \ldots$

(ii) $1, \frac{1}{2}, \frac{1}{3}, \frac{1}{4}, \ldots \frac{1}{n}, \ldots$

(iii) $1, 1, 1, 2, 2, 2, 2, 2, 3, 3, \ldots [\sqrt{n}], \ldots$

(where the nth term is the integer part of the square root of n) and

(iv) $0.841\ldots, 0.909\ldots, 0.141\ldots, -0.756\ldots, \ldots \sin n, \ldots$

(where 'sin' need mean nothing more to us at the moment than a way of getting the terms of a sequence by pushing a button on a calculator – set to radians in this instance).

The graphs of these sequences are illustrated below:

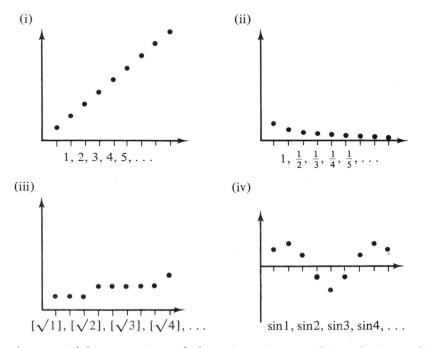

(i)

1, 2, 3, 4, 5, . . .

(ii)

$1, \frac{1}{2}, \frac{1}{3}, \frac{1}{4}, \frac{1}{5}, \ldots$

(iii)

$[\sqrt{1}], [\sqrt{2}], [\sqrt{3}], [\sqrt{4}], \ldots$

(iv)

sin 1, sin 2, sin 3, sin 4, . . .

As you might guess, two of these sequences are 'increasing', one is 'decreasing' and the other is doing neither. In general the sequence $x_1, x_2, x_3, \ldots$ is said to be *increasing* if $x_1 \leqslant x_2 \leqslant x_3 \leqslant \ldots$ and *decreasing* if $x_1 \geqslant x_2 \geqslant x_3 \geqslant \ldots$. Note that the sequence

$$1, 1, 1, \ldots 1, \ldots$$

is both increasing and decreasing, so these terms are really misnomers for 'never decreasing' and 'never increasing'.

The above sequence

sin 1, **sin 2**, sin 3, sin 4, sin 5, sin 6, sin 7, **sin 8**, sin 9, sin 10, . . .

is neither increasing nor decreasing and yet if you calculate the highlighted terms you will see that *they* are increasing (and it's possible to continue to pick out a never-ending list of increasing terms):

sin 1, sin 2, sin 8, sin 14, sin 102, sin 165, . . .

Given a sequence

$$x_1, x_2, x_3, \ldots x_n, \ldots$$

by a *subsequence* we mean another sequence

$$x_{k_1}, x_{k_2}, x_{k_3}, \ldots x_{k_n}, \ldots$$

picked out in order from the original sequence. So for example there is a subsequence

sin 1, sin 2, sin 8, sin 14, sin 102, sin 165, ...

picked out from the sequence

sin 1, sin 2, sin 3, sin 4, sin 5, ...

which is increasing.

Surprisingly it will always be the case that any sequence, no matter how randomly chosen, will have either an increasing subsequence or a decreasing subsequence, as we now prove.

> **Theorem**   Given any sequence either there will exist a subsequence which is increasing or there will exist a subsequence which is decreasing (and possibly both).

Waterproof   Let $x_1, x_2, x_3, \ldots$ be any sequence and consider its graph. For the purposes of illustration assume for the moment that all the x_n are positive.

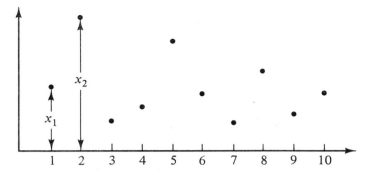

Now assume that the points in that figure represent people on the roofs of their Spanish sky-scraper hotels on the Costa Bom. The hotels stretch in a long line towards the sea in the distance:

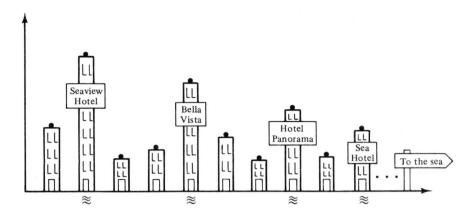

The Thomson representative who is compiling their new brochure visits the resort and gives each hotel with a sea-view a special symbol ≈ in the brochure. So a hotel gets the ≈ symbol if a person standing on the roof and looking towards the sea does not have his view obscured by some taller hotel nearer the sea. Those hotels which earned the special symbol are shown in the picture above.

Now one of two statements will be true about this resort:

Either there will be infinitely many hotels with the ≈ symbol.	**or** there is only a finite number of hotels with the ≈ symbol.
In this case let the hotels with symbol ≈ be numbered $k_1, k_2, k_3, \ldots$	In this case there will be a last hotel with the symbol: let the first hotel after that be the one numbered k_1. (If there are no hotels with the ≈ symbol we'll take k_1 to be 1.)

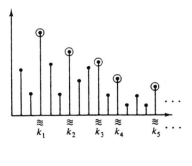

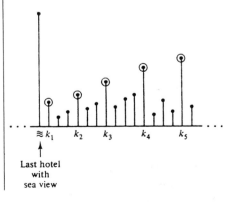

Last hotel
with
sea view

Then, as none of these hotels has its view obscured by any higher hotel nearer the sea it follows that:

$$x_{k_1} > x_{k_2} > x_{k_3} > \ldots$$

and we have found a decreasing subsequence.

Then the hotel k_1 has its view obscured by some higher hotel, number k_2 say. This hotel doesn't have a sea-view either so its view of the sea must be blocked by some higher hotel, number k_3 say. In this way we get an increasing subsequence

$$x_{k_1} \leqslant x_{k_2} \leqslant x_{k_3} \leqslant \ldots$$

So in each case we have found an increasing or a decreasing subsequence as required.   □

Purists may not like our watery proof but the ideas there can be made exact, as in the following formal interpretation of the above proof.

Water-tight proof   Given a sequence $x_1, x_2, x_3, \ldots$ let the set C (which can be thought of as *see* the *sea*) of positive integers be given by

$$C = \{N \in \mathbb{N}: \text{if } m > N \text{ then } x_m < x_N\}$$

Either C is not bounded above.

or C is bounded above.

In this case there exist

$$k_1 < k_2 < k_3 < \cdots$$

in C. The fact that $k_1 \in C$ ensures that if $m > k_1$ then $x_m < x_{k_1}$. In particular $x_{k_2} < x_{k_1}$. Then the fact that $k_2 \in C$ ensures that if $m > k_2$ then $x_m < x_{k_2}$. In particular $x_{k_3} < x_{k_2}$. Continuing in this way we can see that

$$x_{k_1} > x_{k_2} > x_{k_3} > \cdots$$

and we have found a decreasing subsequence.

In this case either C is empty or it has a least upper bound. Hence there exists a positive integer k_1 such that none of $k_1, k_1 + 1, k_1 + 2, \ldots$ are in C. But as $k_1 \notin C$ there must exist an integer m with $m > k_1$ and $x_m \geqslant x_{k_1}$: call this integer k_2. Similarly $k_2 \notin C$ and so there exists an integer k_3 with $k_3 > k_2$ and $x_{k_3} \geqslant x_{k_2}$. Continuing in this way yields an increasing subsequence

$$x_{k_1} \leqslant x_{k_2} \leqslant x_{k_3} \leqslant \cdots$$

Hence, as before, in each case we have found an increasing or a decreasing subsequence.   □

Increasing and decreasing sequences will feature in some of our applications. For example consider the function

$$f(x) = \sqrt{x} \quad x \geqslant 0$$

We shall calculate a sequence of approximations to the area under the graph from $x = 1$ to $x = 9$:

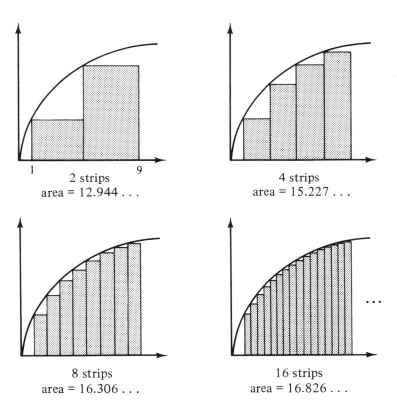

2 strips
area = 12.944 . . .

4 strips
area = 15.227 . . .

8 strips
area = 16.306 . . .

16 strips
area = 16.826 . . .

Here we see that doubling the number of columns leads to a sequence of areas of rectangles which is increasing. (It's fairly easy to see why: in going from one stage to the next each strip is split into two and one of the new narrower rectangles can be higher.)

When, in about 250 BC, Archimedes calculated approximate values of π he used a sequence of regular polygons getting 'closer and closer to a circle'. We illustrate some examples in a circle of radius 1 and you can see again why this sequence of areas is increasing: each is obtained from the former by the addition of some triangles.

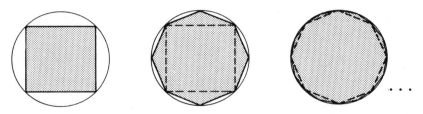

4 sides: area = 2   8 sides: area = 2.828 . . .   16 sides: area = 3.061 . . .

Of course in those two examples not only are the sequences increasing but they are, in some sense, tending towards a particular number. We shall investigate this idea after the next exercises.

Exercises

1   Show that in the sequence
$$x_1 = 1, x_2 = 1 - \tfrac{1}{2}, x_3 = 1 - \tfrac{1}{2} + \tfrac{1}{4}, x_4 = 1 - \tfrac{1}{2} + \tfrac{1}{4} - \tfrac{1}{8}, \ldots$$
the subsequence $x_1, x_3, x_5, \ldots$ is decreasing and that the subsequence $x_2, x_4, x_6, \ldots$ is increasing. Show also that each even-numbered term is less than each odd-numbered term.

Draw a rough sketch of the graph of the sequence.

By using the formula for the sum of the geometric progression
$$1 - \tfrac{1}{2} + \tfrac{1}{4} - \tfrac{1}{8} + \cdots \pm \frac{1}{2^{n-1}}$$

make a guess at which number you think the sequence is tending towards.

2   For each integer n let x_n denote the area of the triangle shown.

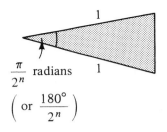

$\dfrac{\pi}{2^n}$ radians

$\left(\text{or } \dfrac{180°}{2^n} \right)$

Without actually calculating numerical values of the x_ns explain informally why the sequence $x_1, x_2, x_3, \ldots$ is decreasing. What number do you think the sequence is tending towards?

Explain similarly why the sequence
$$x_1, 2x_2, 4x_3, 8x_4, \ldots$$

is increasing. Is it tending towards any particular number and, if so, what do you think that number is?

3   Let f be a function whose domain is an interval. Then f is called *convex* if whenever a chord is drawn between two points on its graph that chord lies either on or above the graph, as illustrated in the left-hand figure.

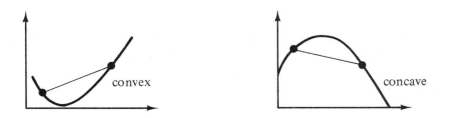

Earlier we considered the convex function given by $f(x) = x^2$ $(x \in \mathbb{R})$ and calculated a sequence of gradients of chords from P $(1, 1)$ to other points on the graph of f. These points were to the right of P and closer and closer to it. The sequence we obtained was

$$2.8, 2.4, 2.2, 2.1, 2.05, \ldots$$

which is decreasing. We shall now see that a sequence obtained in this way for a convex function will always be decreasing.

So now let f be any convex function and consider the points P $(x_0, f(x_0))$ and Q $(x_1, f(x_1))$ on the graph of f. Let x_2 satisfy $x_0 < x_2 < x_1$. Find the coordinates of the point on the chord PQ whose x-coordinate is x_2 and deduce that

$$f(x_2) \leqslant \frac{f(x_0)(x_1 - x_2) + f(x_1)(x_2 - x_0)}{x_1 - x_0}$$

Deduce further that if Q' is the point $(x_2, f(x_2))$ on the graph of f then

gradient of $PQ' \leqslant$ gradient of PQ

Hence show that a sequence of gradients of chords of f from P to points closer and closer to P (and to the right of it) will be decreasing.

4   A *concave* function is one whose domain is an interval and is such that any of its chords lies on or below the graph of the function, as illustrated on the right above. By using a mirror or otherwise show that if a sequence of gradients is obtained exactly as in the previous exercise, then the sequence will be increasing.

5   In exercise 6 on page 19 we established Bernoulli's inequality,

namely that $(1 + x)^n \geqslant 1 + nx$ for $x \geqslant -1$. Use this inequality with $x = 1/n^2$ to show that for $n > 1$

$$\left(1 + \frac{1}{n}\right)^n \geqslant \frac{1}{(1 - 1/n)^{n-1}} = \left(1 + \frac{1}{n-1}\right)^{n-1}$$

Deduce that the following sequence is increasing:

$$2, (1\tfrac{1}{2})^2, (1\tfrac{1}{3})^3, (1\tfrac{1}{4})^4, \ldots \left(1 + \frac{1}{n}\right)^n, \ldots$$

Keen readers could use the binomial theorem's expansion of $(1 + x)^n$ with $x = 1/n$ to deduce that each term of this sequence is less than e.

By calculating some terms of this sequence see if you can guess which number (if any) it is 'tending towards'. (Much later in the book – exercise 5 on page 168 – we shall actually prove this and much more.)

(Solutions on page 233)

Tending towards ...?

One of the most difficult concepts encountered in a first-year maths course at college is that of 'convergence' where we wish to make precise the idea of a sequence 'tending towards' some number. We have already used this idea informally several times. For example in some of our earlier work on gradients we said that the sequence

$$2.8, 2.4, 2.2, 2.1, 2.05, \ldots$$

seemed to be 'tending towards' 2, and with our calculators we constructed a sequence

$$10, 5.1, 2.746078, 1.737194, 1.444238, 1.414525, 1.414213, \ldots$$

apparently 'tending towards' $\sqrt{2}$. But what do we really mean by a sequence 'tending towards' some number?

It's slightly easier to start by considering sequences which 'tend to 0': which of the following sequences do you think have that property?

(i) $1, \frac{1}{2}, \frac{1}{3}, \frac{1}{4}, \frac{1}{5}, \frac{1}{6}, \ldots$

(ii) $1, -1, 1, -1, 1, -1, \ldots$

(iii) $1, -\dfrac{1}{\sqrt{2}}, \dfrac{1}{\sqrt{3}}, -\dfrac{1}{\sqrt{4}}, \dfrac{1}{\sqrt{5}}, -\dfrac{1}{\sqrt{6}}, \ldots$

(iv) $1, \frac{1}{2}, \frac{1}{3}, 1, \frac{1}{4}, \frac{1}{5}, 1\frac{1}{6}, \frac{1}{7}, 1, \ldots$

(v) $1, 0, \frac{1}{2}, 0, \frac{1}{3}, 0, \frac{1}{4}, 0, \ldots$

From those sequences the ones which it seems sensible to say are 'tending towards 0' are (i) (iii) and (v). However close you want to get to 0 these sequences eventually get that close and stay that close. For example let us consider (iii)

$$1, \ -\frac{1}{\sqrt{2}}, \frac{1}{\sqrt{3}}, \ -\frac{1}{\sqrt{4}}, \frac{1}{\sqrt{5}}, \ -\frac{1}{\sqrt{6}}, \cdots$$

If you said

'Convince me that the terms here eventually get within 0.01 of 0 and stay that close,'

then I'd say

'Look at the terms from the 10 001st onwards: they are

$$\frac{1}{\sqrt{10\,001}}, \ -\frac{1}{\sqrt{10\,002}}, \frac{1}{\sqrt{10\,003}}, \ -\frac{1}{\sqrt{10\,004}}, \cdots$$

(≈ 0.0099)

all of which *are* within 0.01 of 0'.

You might then try to make my task harder by saying

'Convince me that the terms eventually get within 0.0004 of 0 and stay that close.'

My reply would be

'Look at the terms from the 6 250 001st onwards. They are

$$\frac{1}{\sqrt{6\,250\,001}}, \ -\frac{1}{\sqrt{6\,250\,002}}, \frac{1}{\sqrt{6\,250\,003}}, \ -\frac{1}{\sqrt{6\,250\,004}}, \cdots$$

($\approx 0.000\,399$)

all of which *are* within 0.0004 of 0'.

Clearly we could go on like this for a very long time. In general you challenge me with a very small positive number (let us call it ε) and I have to produce some term in the sequence (let us call it the Nth) such that the Nth, $(N + 1)$th, $(N + 2)$th, ... and all subsequent terms are within ε of 0.

Example Consider the sequence
$$1, \ -\tfrac{1}{2}, \tfrac{1}{3}, \ -\tfrac{1}{4}, \tfrac{1}{5}, \ldots$$
(i) Find an N such that the Nth terms onwards are all within 0.01 of 0 (i.e. are all between -0.01 and $+0.01$).
(ii) Find an N such that the Nth terms onwards are all within 0.0005 of 0 (i.e. are all between -0.0005 and $+0.0005$).

(iii) Show that if ε is any positive number (no matter how small) and N is any integer larger than $1/\varepsilon$ then the Nth terms onwards are all within ε of 0 (i.e. are all between $-\varepsilon$ and $+\varepsilon$).

Solution (i) Take $N = 101$. Then the 101st, 102nd, 103rd terms of the sequence are

$$\frac{1}{101}\,(=0.0099\ldots),\quad -\frac{1}{102}\,(=-0.0098\ldots),\quad \frac{1}{103}\,(=0.0097\ldots),\ldots$$

all of which *are* within 0.01 of 0.

(ii) Take $N = 2001$. Then the 2001st, 2002nd, 2003rd, ... terms are

$$\frac{1}{2001}\,(=0.0004997\ldots),\quad -\frac{1}{2002}\,(=-0.0004995\ldots),$$

$$\frac{1}{2003}\,(-0.0004992\ldots),\ldots$$

all of which *are* within 0.0005 of 0.

(iii) In general if ε is any positive number and N is an integer larger than $1/\varepsilon$ then

$$-\varepsilon < \pm\frac{1}{N} < \varepsilon,\qquad -\varepsilon < \pm\frac{1}{N+1} < \varepsilon,$$

$$-\varepsilon < \pm\frac{1}{N+2} < \varepsilon,\ldots$$

so that the terms from the Nth onwards are all within ε of 0 as required.

(Actually in this case the terms get steadily closer to 0 and therefore once the Nth term is within ε of 0 so will all the subsequent terms be.) □

In general to show that a sequence $x_1, x_2, x_3, \ldots$ 'tends to' some number x (called the 'limit') the sequence will have to satisfy the following stringent test:

> Whenever you give me a positive number ε (no matter how small) I must be able to find some term of the sequence (the Nth say) such that it and all subsequent terms are within ε of x; i.e. such that all of $x_N, x_{N+1}, x_{N+2}, \ldots$ lie between $x - \varepsilon$ and $x + \varepsilon$.

Example Show that the sequence
$$\tfrac{1}{2}, \tfrac{1}{2} + \tfrac{1}{4}, \tfrac{1}{2} + \tfrac{1}{4} + \tfrac{1}{8}, \tfrac{1}{2} + \tfrac{1}{4} + \tfrac{1}{8} + \tfrac{1}{16}, \ldots$$
satisfies the above test and 'tends to' 1.

Solution   By using the sum of a geometric progression we can see that the nth term of this sequence is

$$x_n = \tfrac{1}{2} + \tfrac{1}{4} + \tfrac{1}{8} + \tfrac{1}{16} + \cdots + \frac{1}{2^n} = 1 - \frac{1}{2^n}$$

Hence the difference between this nth term and our hoped-for limit 1 is $1/2^n$. To make sure that from the Nth term onwards this difference is less than ε we can choose N to be any integer larger than $1/\varepsilon$. For then

$$\frac{1}{2^N} < \frac{1}{N} < \varepsilon \quad \text{and} \quad \frac{1}{2^{N+1}} < \frac{1}{2^N} < \varepsilon \quad \text{and} \quad \frac{1}{2^{N+2}} < \varepsilon \quad \text{and} \ldots$$

(Again in this case the terms are getting steadily closer to 1, so once the Nth term is within ε of 1 all the subsequent terms will also be within ε of 1.)

Hence we have found an N with

$$1 - \varepsilon < 1 - \frac{1}{2^N} = x_N < 1 + \varepsilon \text{ and } 1 - \varepsilon < x_{N+1} < 1 + \varepsilon$$

and $1 - \varepsilon < x_{N+1} < 1 + \varepsilon$ and $\ldots$

and the sequence does satisfy the new formal test and 'tends to' 1.   □

We are now nearly ready for the formal definition. A sequence $x_1, x_2, x_3, \ldots$ will 'tend to' or 'converge to' a 'limit' x if given any positive number ε (no matter how small) there exists an integer N such that

$x - \varepsilon < x_N < x + \varepsilon$ and $x - \varepsilon < x_{N+1} < x + \varepsilon$
and $x - \varepsilon < x_{N+2} < x + \varepsilon$ and $\ldots$

We have defined many new terms already ('rational', 'bounded' etc) but the definition of convergence is so fundamental to a development of analysis that we display it for future reference:

> **Definition**   The sequence $x_1, x_2, x_3, \ldots$ *converges to* the *limit* x (or *tends to* x) if given any $\varepsilon > 0$ there exists an integer N with
> $x - \varepsilon < x_N < x + \varepsilon$ and $x - \varepsilon < x_{N+1} < x + \varepsilon$ and
> $x - \varepsilon < x_{N+2} < x + \varepsilon$ and $\ldots$
> We then write '$x_1, x_2, x_3, \ldots \to x$' (or '$x_n \to x$ as $n \to \infty$' or '$\lim x_n = x$').
>
> A sequence is called *convergent* if such an x exists: otherwise it is called *divergent*.

Before proceeding to some exercises on that definition we make several comments. Firstly convergence concerns the behaviour of the 'tail-end' of a sequence and changing the first 100 terms (or the first 1000, or the first 1 000 000) will not affect the convergence of a sequence. Secondly, most text-books abbreviate the inequalities

$$x - \varepsilon < x_n < x + \varepsilon \text{ to } |x_n - x| < \varepsilon$$

These are equivalent because

$$x - \varepsilon < x_n < x + \varepsilon \underset{\text{'if and only if'}}{\Leftrightarrow} -\varepsilon < x_n - x < \varepsilon \Leftrightarrow \underbrace{|x_n - x| < \varepsilon}_{\substack{\text{the numerical} \\ \text{size of } x_n - x}}$$

Finally, note that we shall soon develop some new techniques and short-cuts for deriving the convergence of a sequence. The formal test, as stated in that definition, will (luckily) only rarely be used. But at least we now have a strict understanding of a sequence tending to a limit and if ever we are in doubt about a particular sequence we can resort to the formal definition.

Exercises

1   Use the definition to show formally that the sequence

$$1, \tfrac{2}{3}, \tfrac{3}{5}, \tfrac{4}{7}, \tfrac{5}{9}, \dots$$

converges to the limit $\tfrac{1}{2}$.

2   Recall that (for $x > 0$) $\log_{10} x = y$ means that $10^y = x$. Use the definition of convergence to show formally that the sequence

$$\frac{1}{\log_{10} 2}, \frac{1}{\log_{10} 3}, \frac{1}{\log_{10} 4}, \frac{1}{\log_{10} 5}, \dots$$

converges to 0.

3   Write down the definition of the sequence $x_1, x_2, x_3, \dots$ converging to x, and also write down the same definition applied to the sequence $x_1 - x, x_2 - x, x_3 - x, \dots$ converging to 0. Observe that the two statements are saying equivalent things. (Hence $x_1, x_2, x_3, \dots$ converges to x if and only if $x_1 - x, x_2 - x, x_3 - x, \dots$ converges to 0.)

Deduce that if the sequence $x_1, x_2, x_3, \dots$ converges to x then for any number y the sequence $x_1 + y, x_2 + y, x_3 + y, \dots$ converges to $x + y$ (i.e. you can 'add a constant to a convergent sequence').

4   By a similar process to that in exercise 3 prove that the sequence $x_1, x_2, x_3, \dots$ converges to x if and only if the sequence $-x_1, -x_2, -x_3, \dots$ converges to $-x$.

5   Prove that the sequence $x_1, x_2, x_3, \dots$ converges to 0 if and only if the sequence $|x_1|, |x_2|, |x_3|, \dots$ converges to 0.

Give an example of a sequence $x_1, x_2, x_3, \ldots$ which is divergent but is such that the sequence $|x_1|, |x_2|, |x_3|, \ldots$ is convergent.

6   Assume that the sequence $x_1, x_2, x_3, \ldots$ of non-negative numbers converges to 0 and that the sequence $y_1, y_2, y_3, \ldots$ satisfies $-x_n \leqslant y_n \leqslant x_n$ for each n. By the definition of the convergence of $x_1, x_2, x_3, \ldots$, given any $\varepsilon > 0$ there exists an integer N such that $x_N, x_{N+1}, x_{N+2}, \ldots$ have a certain property: show that $y_N, y_{N+1}, y_{N+2}, \ldots$ have the same property and deduce that the sequence $y_1, y_2, y_3, \ldots$ also converges to 0.

(Solutions on page 236)

Bound to get there

We have talked about *the* limit of a convergent sequence, but to be able to refer to such a thing unambiguously we ought to check that a convergent sequence cannot have two different limits:

> **Theorem**   If $x_1, x_2, x_3, \ldots \to x$ and $x_1, x_2, x_3, \ldots \to y$, then $x = y$; i.e. a convergent sequence has a unique limit.

> **Proof**   We shall assume that the sequence has two different limits x and y with $x < y$ say, and we shall derive a contradiction. Since $x < y$ we can apply the definition of convergence with ε taken as the positive number $\frac{1}{2}(y - x)$.

By considering $x_1, x_2, x_3, \ldots \to x$ we deduce that

$$x - \tfrac{1}{2}(y - x) < x_n < x + \tfrac{1}{2}(y - x)$$

for all n bigger than or equal to some integer N_1.

By considering $x_1, x_2, x_3, \ldots \to y$ we deduce that

$$y - \tfrac{1}{2}(y - x) < x_n < y + \tfrac{1}{2}(y - x)$$

for all n bigger than or equal to some integer N_2.

Now let N be any integer larger than N_1 and larger than N_2. Then in particular

$$x_N < x + \tfrac{1}{2}(y - x) \quad \text{and} \quad y - \tfrac{1}{2}(y - x) < x_N$$

But then

$$\tfrac{1}{2}(x + y) = y - \tfrac{1}{2}(y - x) < x_N < x + \tfrac{1}{2}(y - x) = \tfrac{1}{2}(x + y)$$

So from our assumption of two different limits we have deduced the obvious contradiction $\frac{1}{2}(x + y) < \frac{1}{2}(x + y)$! Hence a convergent sequence has just one limit.   □

Naturally enough a sequence $x_1, x_2, x_3, \ldots$ is said to be *bounded above* if the set of its terms $\{x_1, x_2, x_3, \ldots\}$ is bounded above. In a similar way we can extend the notions of 'bounded below' and 'bounded' from sets to sequences. It is quite possible for sequences to be unbounded (i.e. not bounded): e.g.

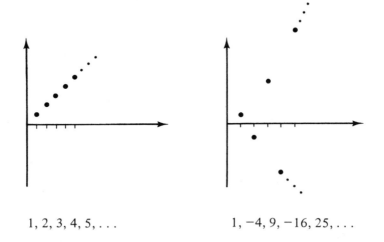

1, 2, 3, 4, 5, . . .   1, −4, 9, −16, 25, . . .

and it's quite possible for a bounded sequence to be divergent: e.g.

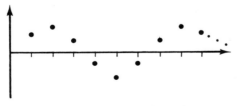

sin 1, sin 2, sin 3, sin 4, sin 5, . . .

(The formal verification that this example diverges will be found as a later exercise.)

However we now see that a convergent sequence will certainly be bounded because after its first few terms a convergent sequence is very restricted indeed.

Theorem   A convergent sequence is bounded.

Proof   Let $x_1, x_2, x_3, \ldots$ be a sequence which converges with limit x. Then from the definition of convergence with ε chosen to be 1 we know that there exists an N with

$$x - 1 < x_N < x + 1, \quad x - 1 < x_{N+1} < x + 1, \quad x - 1 < x_{N+2} < x + 1, \dots$$

i.e. all of the terms $x_N, x_{N+1}, x_{N+2}, \dots$ lie between $x - 1$ and $x + 1$. Now split the set $\{x_1, x_2, x_3, \dots\}$ into two parts:

$$\{x_1, x_2, x_3, \dots\} = \underbrace{\{x_1, x_2, \dots, x_{N-1}\}}_{\substack{\text{a finite set} \\ \therefore \text{ bounded}}} \cup \underbrace{\{x_N, x_{N+1}, x_{N+2}, \dots\}}_{\substack{\text{bounded between} \\ x - 1 \text{ and } x + 1}}$$

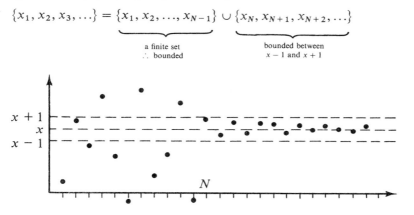

Hence $\{x_1, x_2, x_3, \dots\}$ is the union of two bounded sets and is itself bounded, as required. □

We have met several examples of increasing sequences; e.g.

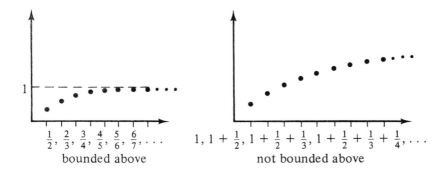

$\dfrac{1}{2}, \dfrac{2}{3}, \dfrac{3}{4}, \dfrac{4}{5}, \dfrac{5}{6}, \dfrac{6}{7}, \dots$

bounded above

$1, 1 + \dfrac{1}{2}, 1 + \dfrac{1}{2} + \dfrac{1}{3}, 1 + \dfrac{1}{2} + \dfrac{1}{3} + \dfrac{1}{4}, \dots$

not bounded above

Clearly a sequence which is not bounded above cannot converge (because we have proved that a convergent sequence must be bounded). But the left-hand sequence illustrated, which is increasing and bounded above, certainly is convergent and from the picture of its graph it seems that its limit is the lowest of all the upper bounds of the sequence. This will always be the case, as we shall soon prove. Before proceeding to that proof I'd like to comment that analysis proofs involving εs are not very readable, but they *are* absolutely logical and a natural consequence of the definitions. Therefore it is much more instructive for the reader to think about the proofs for him or herself than to just read them. For this reason in the next few proofs I have left some very straightforward gaps, denoted by empty boxes, for you to fill in.

> **Theorem**   Let $x_1, x_2, x_3, \dots$ be an increasing sequence which is bounded above. Then it is convergent and its limit is $\sup\{x_1, x_2, x_3, \dots\}$.

> **Proof**   The set $\{x_1, x_2, x_3, \dots\}$ is non-empty and bounded above and so it has a least upper bound or supremum, α say. We aim to show that $x_1, x_2, x_3, \dots \to \alpha$.

Let ε be any positive number. By the definition of convergence we must find an N with

$$\alpha - \varepsilon < \boxed{} < \alpha + \varepsilon, \quad \alpha - \varepsilon < \boxed{} < \alpha + \varepsilon, \quad \alpha - \varepsilon < \boxed{} < \alpha + \varepsilon, \dots$$

(Half the inequalities come free: because α is an upper bound of the sequence the right-hand inequality is automatically true in each case.)

Now $\alpha - \varepsilon$ is less than α, and α is the $\boxed{}$ of all the upper bounds of the set $\{x_1, x_2, x_3, \dots\}$ and so it follows that $\alpha - \varepsilon$ is *not* an $\boxed{}$ $\boxed{}$ of the set $\{x_1, x_2, x_3, \dots\}$. Hence there exists some x_N with $x_N > \alpha - \varepsilon$. But the sequence is increasing and so we can deduce that

$$\alpha - \varepsilon < x_N \leqslant x_{N+1} \leqslant x_{N+2} \leqslant \cdots \leqslant \alpha < \alpha + \varepsilon$$

Hence

$$\alpha - \varepsilon < x_N < \alpha + \varepsilon, \quad \alpha - \varepsilon < x_{N+1} < \alpha + \varepsilon, \quad \alpha - \varepsilon < x_{N+2} < \alpha + \varepsilon, \dots$$

(or, more succinctly, $\alpha - \varepsilon < x_n < \alpha + \varepsilon$ for all $n \geqslant N$) and we have found the required N.

We have therefore shown that $x_1, x_2, x_3, \dots \to \alpha$ as required.   □

The corresponding result that a decreasing sequence which is bounded below converges to its infimum can be proved similarly, and this is left as one of the next exercises.

> **Example**   We considered earlier a way of calculating the area under the graph of $g(x) = 1/x$ from $x = 1$ to $x = t$ say by obtaining a sequence of approximations to that area. The sequence which we obtained was increasing and bounded above (and hence

convergent) and the limit of that sequence is the area under the graph: we shall denote it in this case by $\log t$ (or $\ln t$). (The verification that this is a sensible definition of area, and the reason for using a 'logarithmic' title will be given in later chapters.)

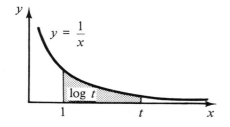

Now consider the sequence $x_2, x_3, x_4, \ldots$ given by

$$x_n = 1 + \tfrac{1}{2} + \tfrac{1}{3} + \cdots + \frac{1}{n-1} - \log n$$

Show that the sequence (x_n) is increasing and bounded above. (Hence (x_n) converges: its limit is called *Euler's constant* and is approximately 0.577: its value was first calculated by the great Swiss mathematician Leonard Euler in the mid-eighteenth century.)

Solution   The number $1 + \tfrac{1}{2} + \tfrac{1}{3} + \cdots + 1/(n-1)$ is the area of the rectangles shown in the left-hand figure below. The number $\log n$ is the area under the curve from $x = 1$ to $x = n$. So x_n, which is the difference of those two, is equal to the area shaded in the right-hand figure.

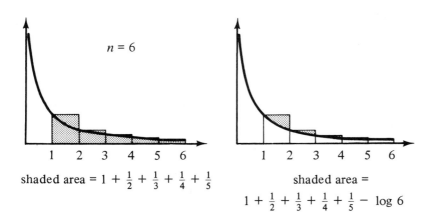

shaded area $= 1 + \tfrac{1}{2} + \tfrac{1}{3} + \tfrac{1}{4} + \tfrac{1}{5}$

shaded area $=$

$1 + \tfrac{1}{2} + \tfrac{1}{3} + \tfrac{1}{4} + \tfrac{1}{5} - \log 6$

Therefore the sequence is clearly increasing because x_{n+1} equals x_n plus an additional area between $x = n$ and $x = n + 1$. The sequence is also bounded above by 1 since the shaded area equalling x_n fits into a 1×1 square as shown.

Hence the sequence (x_n) is increasing and bounded above, and is therefore convergent. (My computer gives the 1000th and 10 000th terms as 0.576 716 and 0.577 166.)   □

We have seen that

convergent $\underset{\text{implies}}{\Rightarrow}$ bounded

bounded and increasing $\Rightarrow$ convergent

and we have noted that in general

bounded $\not\Rightarrow$ convergent

as with the sequence sin 1, sin 2, sin 3, ... However we saw informally on page 31 that this sequence has an increasing subsequence

sin 1, sin 2, sin 8, sin 14, sin 102, sin 165, ...

This subsequence is therefore increasing and bounded above (by 1 for example). It follows from the above theorem that the subsequence is convergent. A similar principle can be applied in general to enable us to deduce the following surprising result:

Theorem Any bounded sequence has a convergent subsequence.

Proof Consider any bounded sequence. Then by the 'Spanish hotel' result on page 32 the sequence either has an [   ] subsequence or else it has a [   ] subsequence. But our sequence here is bounded. So we have either

an [   ] subsequence which is [   ] [a   ]

or

a [   ] subsequence which is [   ] [b   ]

By the previous theorem (and by the comment after it concerning the decreasing case) this subsequence must be convergent, and the theorem is proved.   □

What about those sequences which are increasing but not bounded above, like

$$1, 2, 3, 4, 5, \ldots \quad \text{and} \quad 1, 4, 9, 16, 25, \ldots ?$$

You might be tempted to say that they 'tend to ∞'. (Let me remind you one more time that ∞ is not a number: we are merely going to try to make some sense of the phrase 'tends to ∞'.)

Increasing sequences are not the only ones which we'd expect to 'tend to ∞'. Surely the sequence

$$\sqrt{2}, \sqrt{1}, \sqrt{4}, \sqrt{3}, \sqrt{6}, \sqrt{5}, \sqrt{8}, \sqrt{7}, \sqrt{10}, \ldots, \sqrt{1000}, \sqrt{999}, \sqrt{1002}, \ldots$$

also has that property?

When we formally introduced the idea of a sequence converging to a limit x you gave me any positive number and, no matter how small it was, I had to convince you that the terms of the sequence eventually got that close and stayed that close to x. A similar analytical principle will enable us to define the idea of 'tending to ∞'.

Consider as an example the sequence

$$\sqrt{2}, \sqrt{1}, \sqrt{4}, \sqrt{3}, \sqrt{6}, \sqrt{5}, \sqrt{8}, \sqrt{7}, \ldots$$

You might say to me 'convince me that those terms eventually all exceed 100', to which my reply would be that the 10 001st, 10 002nd, 10 003rd, terms etc are $\sqrt{10\,002}$ (≈ 100.01), $\sqrt{10\,001}$ (≈ 100.005), $\sqrt{10\,004}$ (≈ 100.02), ... all of which *are* larger than 100. Realising that you had made the task too easy you might then ask me to convince you that the terms eventually all exceed 4500. I could then show you that the 20 250 001st terms onwards all exceed 4500; for example the 20 250 001st term itself is $\sqrt{20\,250\,002}$ (≈ 4500.0002). We could go on and on like this – you give me a number k and, no matter how large your choice was, I have to convince you that the terms of the sequence eventually all exceed that k. You've probably got the hang of this analytical repartee by now and you'll be ready for the formal definition.

> ***Definition***   A sequence $x_1, x_2, x_3, \ldots$ is said to *tend to infinity* (or *tend to ∞*) if given any number k there exists an integer N with

$x_N > k$, $x_{N+1} > k$, $x_{N+2} > k, \ldots$ (or, more succinctly, $x_n > k$ for all $n \geqslant N$). We write '$x_1, x_2, x_3, \ldots \to \infty$' (or '$x_n \to \infty$ as $n \to \infty$' or '$\lim x_n = \infty$').

The sequence $x_1, x_2, x_3, \ldots$ *tends to* $-\infty$ if $-x_1, -x_2, -x_3, \ldots$ tends to ∞.

Example Show that the sequence $\log_{10} 1$, $\log_{10} 2$, $\log_{10} 3$, $\log_{10} 4, \ldots$ tends to ∞.

Solution The nth term of the sequence is given by $x_n = \log_{10} n$. We have to show that $x_1, x_2, x_3, \ldots$ tends to ∞.

Given any number k (no matter how huge it is) I have to try to find an integer N such that $x_N, x_{N+1}, x_{N+2}, \ldots$ are all bigger than k. But $x_n = \log_{10} n$ and this will be bigger than k when n is bigger than 10^k. So let N be any integer larger than 10^k. Then for any $n \geqslant N$ we have

$$x_n = \log_{10} n \geqslant \log_{10} N > \log_{10}(10^k) = k$$

as required. We have thus found the required N and shown that the sequence tends to ∞.   □

Exercises

1   Show that if $x_1, x_2, x_3, \ldots$ converges to x then any of its subsequences also converges to x.

Deduce that the sequence

$$1, \tfrac{1}{2}, \tfrac{1}{3}, 1, \tfrac{1}{4}, \tfrac{1}{5}, 1, \tfrac{1}{6}, \tfrac{1}{7}, 1, \ldots$$

is divergent.

2   Show that if $x_1, x_3, x_5, \ldots$ converges to x and $x_2, x_4, x_6, \ldots$ converges to x, then the sequence $x_1, x_2, x_3, x_4, \ldots$ converges to x.

3   Prove that a sequence which is decreasing and bounded below converges.

4   Let $x_1, x_2, x_3, \ldots$ be a sequence of non-negative terms (i.e. $x_n \geqslant 0$ for each n) and assume that the sequence converges to x. (We aim to show that x is also non-negative.) By assuming that $x < 0$ and taking $\varepsilon = -x$ in the definition of convergence, deduce a contradiction and hence show that $x \geqslant 0$.

Deduce that if $y_1, y_2, y_3 \to y$ and each $y_n \geqslant a$, then $y \geqslant a$.

Deduce further that if $z_1, z_2, z_3, \ldots \to z$ and each $z_n \leqslant b$, then $z \leqslant b$. Hence show that any sequence in the closed and bounded interval $[a, b]$ has a subsequence which converges to some number in $[a, b]$.

5   We mentioned earlier that the sequence sin 1, sin 2, sin 3, ... is divergent and now we are going to verify that fact formally. We shall assume that in calculating sin x the x is in radians. So for example $\sin(\pi/6) = \frac{1}{2}$, $\sin(\pi/2) = 1$, $\sin(5\pi/6) = \frac{1}{2}$, $\sin(7\pi/6) = -\frac{1}{2}$, etc., with the graph of the sine function taking its familiar shape:

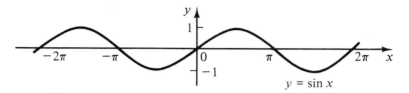

In exercise 3 on page 18 we saw that if two numbers differ by more than 1 then there is an integer between them. Use the fact that $\pi > 3$ to show that there exist integers $k_1, k_2, k_3, \ldots$ with

$$\frac{\pi}{6} < k_1 < \frac{5\pi}{6}, \quad \frac{13\pi}{6} < k_2 < \frac{17\pi}{6}, \quad \frac{25\pi}{6} < k_3 < \frac{29\pi}{6}, \ldots$$

and that there exist integers $j_1, j_2, j_3, \ldots$ with

$$\frac{7\pi}{6} < j_1 < \frac{11\pi}{6}, \quad \frac{19\pi}{6} < j_2 < \frac{23\pi}{6}, \quad \frac{31\pi}{6} < j_3 < \frac{35\pi}{6}, \ldots$$

Deduce that the sequence sin 1, sin 2, sin 3, sin 4, ... has a subsequence of terms all exceeding $\frac{1}{2}$ and another subsequence of terms all less than $-\frac{1}{2}$, and that the sequence is therefore divergent.

6   Let $x_1, x_2, x_3, \ldots$ be a sequence of non-zero terms which converges to the limit x, where $x > 0$. By taking $\varepsilon = x/2$ in the definition of convergence show that there exists an N with $x_N, x_{N+1}, x_{N+2}, \ldots$ all greater than $x/2$. Deduce that the set

$$\left\{ \frac{1}{x_1}, \frac{1}{x_2}, \frac{1}{x_3}, \frac{1}{x_4}, \ldots \right\}$$

is bounded.

7   (i) Let the sequence $x_1, x_2, x_3, \ldots$ of positive terms converge to 0. By taking $\varepsilon = 1$ in the definition of convergence show that there exists an N_1 with $1/x_n > 1$ for all $n \geqslant N_1$.

     By taking $\varepsilon = \frac{1}{2}$ in the definition of convergence show that there exists an N_2 with $1/x_n > 2$ for all $n \geqslant N_2$.

     Prove that the sequence

$$\frac{1}{x_1}, \frac{1}{x_2}, \frac{1}{x_3}, \frac{1}{x_4}, \ldots$$

tends to ∞.

(ii) Let $x_1, x_2, x_3, \ldots$ be a sequence of positive numbers such that

$$\frac{1}{x_1}, \frac{1}{x_2}, \frac{1}{x_3}, \frac{1}{x_4}, \ldots$$

tends to ∞. By taking $k = 1$ in the definition of 'tending to ∞' show that there exists an N_1 with $x_n < 1$ for all $n \geqslant N_1$.

By taking $k = 2$ show that there exists an N_2 with $x_n < \frac{1}{2}$ for all $n \geqslant N_2$. Prove that the sequence $x_1, x_2, x_3, \ldots$ converges to 0.

(Hence a sequence of positive terms converges to 0 if and only if the sequence of their reciprocals tends to infinity.)

(Solutions on page 238)

Some labour-saving devices

This is not quite going to be an ε-free zone but we can gradually develop results about convergence so that we have to resort to εs less and less. For example we proved formally in an earlier exercise that

$$1, \tfrac{2}{3}, \tfrac{3}{5}, \tfrac{4}{7}, \tfrac{5}{9}, \tfrac{6}{11}, \ldots \to \tfrac{1}{2}$$

Then in exercise 3 on page 42 we showed that you can 'add a constant to a convergent sequence'. So for example we can deduce without any further εs that

$$2, 1\tfrac{2}{3}, 1\tfrac{3}{5}, 1\tfrac{4}{7}, 1\tfrac{5}{9}, 1\tfrac{6}{11}, \ldots \to 1\tfrac{1}{2}$$

And surely you would expect the sequence

$$\pi, \tfrac{2}{3}\pi, \tfrac{3}{5}\pi, \tfrac{4}{7}\pi, \tfrac{5}{9}\pi, \tfrac{6}{11}\pi, \ldots$$

to converge to $\pi/2$? We now confirm that you can 'multiply through a convergent sequence by a constant'. (And now that you are becoming familiar with sequences we shall gradually introduce the shorthand notation (x_n) for the sequence $x_1, x_2, x_3, \ldots$.)

> **Theorem** If $x_1, x_2, x_3, \ldots \to x$ then for any number y the sequence $x_1 y, x_2 y, x_3 y, \ldots \to xy$. i.e. if $x_n \to x$ as $n \to \infty$ then $x_n y \to xy$ as $n \to \infty$.)

> **Proof** If $y = 0$ then the result is trivial because it is an immediate consequence of the definition of convergence that $0, 0, 0, \ldots \to 0$. So assume for example that $y > 0$ (the proof can then easily be adapted to the case where $y < 0$). We are given that

$$x_1, x_2, x_3, \ldots \to x$$

and our aim is to show that

$$x_1 y, x_2 y, x_3 y, \ldots \to xy$$

So let ε be any given positive number. We must find an N with

$$xy - \boxed{} < x_n y < xy + \boxed{} \quad \text{for all } n \boxed{}$$

Now y is a positive number and so we can divide through (or multiply through) inequalities by y without upsetting them. Hence N must satisfy

$$x - \frac{\boxed{}}{\boxed{}} < x_n < x + \frac{\boxed{}}{\boxed{}} \quad \text{for all } n \boxed{}$$

But the stringent test outlined in the definition of $x_1, x_2, x_3, \ldots$ converging to x applies to *any* positive number: we now take that positive number to be ε/y. Hence there does exist an N with

$$x - \frac{\varepsilon}{y} < x_n < x + \frac{\varepsilon}{y} \quad \text{for all } n \geqslant N$$

This is exactly what we need because multiplying through these inequalities by y gives

$$\boxed{} < x_n y < \boxed{} \quad \text{for all } \boxed{}$$

Hence given any $\varepsilon > 0$ we have found an N as required to prove that $x_n y \to xy$. $\qquad\qquad\square$

Does your calculator allow you to repeat an operation (with a 'K' button for example)? Choose a number between 0 and 1 and using your calculator produce a sequence of answers obtained by repeatedly multiplying your original number by itself. For example with the number 0.8 I got the sequence

$$0.8, 0.64, 0.512, 0.4096, 0.32768, 0.262144, 0.2097152, 0.1677721, \ldots$$

You will see that the sequence of answers which you are producing seems to be converging to 0, and we can now prove that this will always be the case.

> ***Example*** Let r be any number in the interval $(-1, 1)$ (i.e. r must satisfy $-1 < r < 1$). Show that the sequence (r^n) converges to 0; (i.e. $r, r^2, r^3, \ldots \to 0$).
>
> Deduce that if $r > 1$ the sequence (r^n) tends to infinity.
>
> ***Solution*** To start with assume that $0 \leqslant r < 1$. Then
>
> $$r^{n+1} = r \times r^n \leqslant r^n$$

and so the sequence (r^n) is decreasing. It is also bounded below by 0. Hence (by the result of exercise 3 above) the sequence converges, to α say:

$$r, r^2, r^3, \ldots \to \alpha$$

By the previous theorem we can 'multiply through' this convergent sequence by r to give

$$r \times r, \quad r^2 \times r, \quad r^3 \times r, \ldots \to \alpha \times r$$

i.e.

$$r^2, \quad r^3, \quad r^4, \ldots \to \alpha \times r$$

But we know that this latter sequence converges to α. Hence $\alpha = \alpha r$ or $\alpha(1-r)=0$. Since $1-r$ is not 0 it follows that $\alpha = 0$ and we have shown that in the case when $0 \leqslant r < 1$ the sequence (r^n) converges to 0.

In the case when $-1 < r < 0$ it follows that $0 < -r < 1$ and we know from the above that the sequence $((-r)^n)$ converges to 0 (or $-r, r^2, -r^3, \ldots \to 0$). But we saw in exercise 5 on page 42 that if a sequence converges to 0 the signs of its terms are irrelevant. Hence we can again deduce that (r^n) converges to 0 in this case.

Now if $r > 1$ then $1/r$ is between 0 and 1 and by the above result the sequence $((1/r)^n)$ (which equals $(1/r^n)$) converges to 0. Hence by exercise 7 above the sequence (r^n) tends to infinity in this case. $\square$

We have seen that it is admissible to 'add a constant right through a convergent sequence' and the next step from that is to add (and subtract) two convergent sequences. For example from the fact that

$$1, \tfrac{2}{3}, \tfrac{3}{5}, \tfrac{4}{7}, \ldots, \frac{n}{2n-1}, \ldots \to \tfrac{1}{2} \quad \text{and} \quad \tfrac{1}{2}, \tfrac{2}{3}, \tfrac{3}{4}, \tfrac{4}{5}, \tfrac{5}{6}, \ldots, \frac{n}{n+1}, \ldots \to 1$$

you'd expect their sum

$$1\tfrac{1}{2}, 1\tfrac{1}{3}, 1\tfrac{7}{20}, 1\tfrac{13}{35}, \ldots, \frac{n}{2n-1} + \frac{n}{n+1}, \ldots$$

to converge to $1\tfrac{1}{2}$ and their difference

$$\tfrac{1}{2}, 0, -\tfrac{3}{20}, -\tfrac{8}{35}, \ldots, \frac{n}{2n-1} - \frac{n}{n+1}, \ldots$$

to converge to $-\tfrac{1}{2}$.

Theorem   If the sequence (x_n) converges to x and the sequence (y_n) converges to y then

(i) the sequence $(x_n + y_n)$ converges to $x + y$ (i.e. if $x_1, x_2, x_3, \to x$ and $y_1, y_2, y_3, \ldots \to y$ then $x_1 + y_1, x_2 + y_2, x_3 + y_3, \ldots \to x + y$);

(ii) the sequence $(x_n - y_n)$ converges to $x - y$.

Proof (i) Let $\varepsilon > 0$. We have to find an N with

$$(x + y) - \square < x_n + y_n < \square + \square \quad \text{for all } n \; \square$$

Perhaps if we made sure that the x_ns were within $\frac{1}{2}\varepsilon$ of x and that the y_ns were within $\frac{1}{2}\varepsilon$ of y then their sum would be within ε of $x + y$?

Since $x_1, x_2, x_3, \ldots \to x$ given **any** positive number ($\frac{1}{2}\varepsilon$ for example) we can ensure that the terms of the sequence come that close to x. So there exists an integer N_1 with

$$x - \tfrac{1}{2}\varepsilon < x_n < \square + \square \quad \text{for all } n \geqslant N_1$$

Similarly, by considering the sequence (y_n) we can deduce that there exists an N_2 with

$$y - \square < \square < \square + \square \quad \text{for all } n \; \square$$

Now let N be the $\boxed{}$ of N_1 and N_2, so that for $n \geqslant N$ both the above inequalities work. We can then add the inequalities to deduce that for all $n \geqslant N$

$$(x + y) - \varepsilon = (x - \tfrac{1}{2}\varepsilon) + (y - \tfrac{1}{2}\varepsilon) < x_n + y_n < \square + \square = (x + y) + \varepsilon$$

We have therefore found an N as required to show that

$$x_1 + y_1, x_2 + y_2, x_3 + y_3, \ldots \to x + y$$

(ii) We could prove this result about the difference of sequences by using εs as we did in (i). But we shall gradually wean ourselves away from εs and make easy deductions from earlier results. Here we are given that the sequence (y_n) converges to y. It therefore follows (from exercise 4 on page 42) that the sequence $(-y_n)$ converges to $-y$. But then by (i) we can add the convergent sequences (x_n) and $(-y_n)$ to deduce that the sequence $(x_n - y_n)$ converges to $x - y$ as required.   □

Corollary If $x_n \leqslant y_n \leqslant z_n$ for each positive integer n and the sequences (x_n) and (z_n) both converge to x, then the 'sandwiched' sequence (y_n) also converges to x.

Proof Since $x_n \leqslant y_n \leqslant z_n$ for each n we have

$$0 \leqslant y_n - x_n \leqslant z_n - x_n$$

$$\underbrace{}$$
by the theorem
this $\to x - x = 0$
as $n \to \infty$

$\underbrace{}$
this is even smaller
$\therefore$ by exercise 6 on page 43
this $\to 0$ as $n \to \infty$

Hence as $n \to \infty$ we have $y_n - x_n \to 0$ and by adding convergent sequences

$$y_n = x_n + (y_n - x_n) \to x + 0 = 0$$

as required.   □

We have seen that the sequence (sin n) is divergent. But what about the sequence $((\sin n)/n)$ (i.e. sin 1, $\frac{1}{2}$ sin 2, $\frac{1}{3}$ sin 3, ...)? If you calculate a few terms you will see that they are rapidly shrinking towards 0. (For example the 100th term is approximately -0.005.) Essentially this is because the numbers sin n are bounded (between -1 and 1) and so their ability to upset the convergence of 1, $\frac{1}{2}, \frac{1}{3}, \frac{1}{4}, \ldots$ to 0 is severely limited.

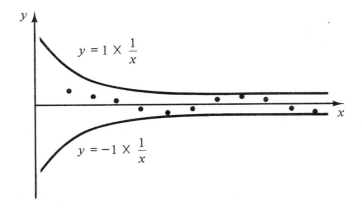

Corollary   If (x_n) is a sequence which converges to 0 and (y_n) is a bounded sequence then the sequence $(x_n y_n)$ also converges to 0.

Proof   Assume that the set $\{y_1, y_2, y_3, \ldots\}$ is bounded below by y and above by z. Then for each positive integer n

$$x_n y \leqslant x_n y_n \leqslant x_n z$$
$$\downarrow \qquad\qquad \downarrow$$
$$0 \times y \qquad\quad 0 \times z$$

Hence we have sandwiched $(x_n y_n)$ between two sequences which converge to 0. It follows from the previous corollary that the sequence $(x_n y_n)$ also converges to 0.   □

We have seen that we can 'add convergent sequences' and in the next corollary we deduce that we can 'multiply convergent sequences':

Corollary   If (x_n) converges to x and (y_n) converges to y then

(i) the sequence $(x_n y_n)$ converges to xy;

(ii) if in addition y and all the y_ns are non-zero then the sequence (x_n/y_n) converges to x/y.

Proof Again this proof is an easy consequence of some properties established earlier:

(i) $x_n y_n = \underbrace{x_n}_{\substack{(x_n) \text{ is convergent} \\ \text{and hence bounded}}} \underbrace{(y_n - y)}_{\substack{\text{this} \to 0 \\ \text{as } n \to \infty}} + \underbrace{x_n y}_{\substack{\text{this} \to xy \\ \text{as } n \to \infty}}$

$\underbrace{\hspace{3cm}}_{\therefore \text{ this} \to 0 \text{ by the theorem}} \qquad \underbrace{\hspace{2cm}}_{\text{this} \to xy}$

$\underbrace{\hspace{7cm}}_{\therefore \text{ this} \to xy \text{ since we can add convergent sequences}}$

Hence the sequence $(x_n y_n)$ converges to xy as required.

(ii) We'll show (with the extra assumptions about none of the ys being 0) that the sequence $(1/y_n)$ converges to $1/y$. The general result about (x_n/y_n) will then follow by applying part (i) to the product of (x_n) and $(1/y_n)$.

$$\frac{1}{y_n} = \frac{1}{y} - \left(\frac{1}{y} - \frac{1}{y_n}\right) = \frac{1}{y} - \underbrace{\frac{1}{yy_n}}_{\substack{\text{these terms are} \\ \text{bounded – exercise 6 page 51}}} \underbrace{(y_n - y)}_{\substack{\text{this} \to 0 \\ \text{as } n \to \infty}}$$

$$\underbrace{\hspace{6cm}}_{\therefore \text{ by the theorem this} \to 0 \text{ as } n \to \infty}$$

$$\underbrace{\hspace{8cm}}_{\therefore \text{ this} \to 1/y \text{ as } n \to \infty}$$

We have thus shown that $1/y_n \to 1/y$ as $n \to \infty$ and, as stated above, the general result that $x_n/y_n \to x/y$ follows from (i). $\qquad \square$

We are now able to add, subtract, multiply, divide and 'sandwich' convergent sequences, enabling us to test the convergence of a wide range of sequences.

Example Show that the sequence $((6n^2 + 1)/(2n^2 + n + 4))$ converges to 3.

(Here the nth terms of the sequence is $(6n^2 + 1)/(2n^2 + n + 1)$, and so the first few terms are $1, \frac{25}{14}, \frac{11}{5}, \ldots$)

Solution As $n \to \infty$ both the numerator and the denominator of the nth term tend to ∞. Since it's virtually impossible to predict the behaviour of such a fraction we 'tame' the numerator and denominator (without affecting the overall value of the fraction) by dividing top and bottom by n^2:

$$n\text{th term} = \frac{6n^2 + 1}{2n^2 + n + 4} = \frac{6 + 1/n^2}{2 + 1/n + 4/n^2}$$

Now we've see that $1/n \to 0$ as $n \to \infty$ and so $1/n^2 \to 0$ as $n \to \infty$ (it's smaller than $1/n$). Hence $4/n^2 \to 0$ as $n \to \infty$ and we see from our results on adding and dividing limits that

$$n\text{th term} = \frac{6 + 1/n^2}{2 + 1/n + 4/n^2} \to \frac{6 + 0}{2 + 0 + 0} = 3$$

as required. □

Example   Calculate the first few terms of $(n^3/2^n)$. You will see that the terms soon start to decrease and seem to tend towards 0. We now aim to show that for any fixed positive integer M and any fixed number $x > 1$ the sequence (n^M/x^n) converges to 0.

(i) Show that there exists an integer N with

$$\frac{(1 + 1/N)^M}{x} < 1$$

(ii) Show that from the Nth term onwards the sequence (n^M/x^n) is decreasing.

(iii) Show that the sequence converges to 0.

Solution   (i) $(1 + 1/n)^1$ converges to 1 as $n \to \infty$. Therefore, multiplying the sequence by itself, $(1 + 1/n)^2$ converges to 1 as $n \to \infty$. Continuing in this way (using induction if you're very fussy) for any fixed positive integer M, $(1 + 1/n)^M$ converges to 1 as $n \to \infty$. Hence

$$\frac{(1 + 1/n)^M}{x} \to \frac{1}{x} < 1 \quad \text{as } n \to \infty$$

So not all the terms of the sequence can be greater than or equal to 1 and there must exist an integer N with

$$\frac{(1 + 1/N)^M}{x} < 1$$

as required.

(ii) For $n \geqslant N$ the $(n + 1)$th term of the sequence divided by the nth term is

$$\frac{(n + 1)^M}{x^{n+1}} \div \frac{n^M}{x^n} = \frac{1}{x}\left(\frac{n + 1}{n}\right)^M = \frac{(1 + 1/n)^M}{x} \leqslant \frac{(1 + 1/N)^M}{x} < 1$$

Hence (except possibly for the first $N - 1$ terms) the sequence is eventually decreasing. It is also bounded below (by 0 for example, since all the terms are positive). Since the first $N - 1$ terms of a sequence do not affect its convergence we deduce that the sequence (n^M/x^n) converges with limit α say.

(iii) As $n \to \infty$

$$\frac{n^M}{x^n} \to \alpha$$

and so

$$\frac{(n+1)^M}{x^{n+1}} = \frac{1}{x} \times \left(1 + \frac{1}{n}\right)^M \times \frac{n^M}{x^n} \to \frac{1}{x} \times 1 \times \alpha = \frac{\alpha}{x}$$

But $(n+1)^M/x^{n+1}$ is the $(n+1)$st term of the sequence and so this too converges to α as $n \to \infty$. Hence $\alpha = \alpha/x$ and it follows that $\alpha = 0$.   □

Example   In your calculator's memory store any positive number x. Calculate the first few terms of the sequence $(x^{1/n})$ (i.e. $x, x^{1/2}, x^{1/3}, \ldots$) and observe that the sequence seems to be converging to 1. (A quick way of seeing this is to display x and then to repeatedly press the 'square root' button. You are then calculating the subsequence

$$x, x^{1/2}, x^{1/4}, x^{1/8}, x^{1/16}, \ldots$$

which rapidly converges to 1.)
Prove that

(i) for any $x \geq 1$ the sequence $(x^{1/n})$ is decreasing and converges to 1;

(ii) for any x with $0 < x \leq 1$ the sequence $(x^{1/n})$ is increasing and converges to 1.

Solution   (i) For $x \geq 1$ we have $x^{n+1} \geq x^n$. Hence, taking $n(n+1)$th roots of both sides of that inequality gives

$$x^{1/n} = (x^{n+1})^{1/n(n+1)} \geq (x^n)^{1/n(n+1)} = x^{1/n+1}$$

It follows that the sequence $(x^{1/n})$ is decreasing. It is also bounded below by 1 (since an nth root of a number greater than 1 must itself be greater than 1). Hence we have a decreasing sequence which is bounded below, and is therefore convergent: assume that its limit is α. Then

$$x, x^{1/2}, x^{1/3}, x^{1/4}, x^{1/5}, \ldots \to \alpha$$

Therefore any subsequence converges to α and in particular we see that

$$x^{1/2}, x^{1/4}, x^{1/6}, x^{1/8}, x^{1/10}, \ldots \to \alpha$$

We know that we can multiply convergent sequences together: multiplying this last convergent sequence by itself gives

$$x^{1/2}x^{1/2}, x^{1/4}x^{1/4}, x^{1/6}x^{1/6}, x^{1/8}x^{1/8}, x^{1/10}x^{1/10}, \ldots \to \alpha^2$$

$$\begin{array}{ccccc} \| & \| & \| & \| & \| \\ x & x^{1/2} & x^{1/3} & x^{1/4} & x^{1/5} \quad \ldots \end{array}$$

But this is just the original sequence $(x^{1/n})$ which converges to α. Hence $\alpha = \alpha^2$ and α must be 0 or 1. Since no term of the sequence is less than 1 it follows that $\alpha = 1$. Hence we have shown that the sequence $(x^{1/n})$ converges to 1.

(ii) For $0 < x \leqslant 1$ we could start again and follow a similar argument. But it is easier to note that $1/x \geqslant 1$. Hence by (i) the sequence $((1/x)^{1/n})$ (which equals $(1/x^{1/n})$) is decreasing and convergent to 1. But by the above corollary we can take reciprocals of convergent sequences: it follows that the sequence $(x^{1/n})$ is increasing and convergent to 1.   □

We now consider a couple of examples where the sequence is not given explicitly but defined by a 'recurrence relation' where each term is obtained from previous terms in some way.

Example   Let the sequence (x_n) be given by
$$x_1 = 2.5 \quad \text{and} \quad x_n = \tfrac{1}{5}(x_{n-1}^2 + 6) \quad \text{for } n > 1$$
Show that

(i) $2 \leqslant x_n \leqslant 3$ for all n;

(ii) (x_n) is decreasing;

(iii) (x_n) converges, and find the limit.

Solution   (i) We prove these inequalities by induction on n, the case $n = 1$ being obvious since we are given that $x_1 = 2.5$. So assume that the inequalities hold for $k - 1$. Then

$$2 \leqslant x_{k-1} \leqslant 3$$
$$\therefore \ 2 = \tfrac{1}{5}(4 + 6) \leqslant \underbrace{\tfrac{1}{5}(x_{k-1}^2 + 6)}_{x_k} \leqslant \tfrac{1}{5}(9 + 6) = 3$$

i.e. $2 \leqslant x_k \leqslant 3$

and we have established the result in the case $n = k$. Hence by induction the inequalities hold for each positive integer n.

(ii) To show that the sequence (x_n) is decreasing we show that $x_{n+1} - x_n \leqslant 0$ for each n:

$$x_{n+1} - x_n = \tfrac{1}{5}(x_n^2 + 6) - x_n = \tfrac{1}{5}(x_n^2 - 5x_n + 6)$$
$$= \tfrac{1}{5}\underbrace{(x_n - 3)}_{\leqslant 0}\underbrace{(x_n - 2)}_{\geqslant 0 \text{ (by(i))}} \leqslant 0$$

Hence $x_{n+1} \leqslant x_n$ for each n and the sequence (x_n) is decreasing.

(iii) The sequence (x_n) is decreasing and bounded below (by 2 for example) and so it converges, with limit x say. Then as $n \to \infty$ we have (using our various results about manipulating convergent sequences)

$$x_{n-1} \to x, \quad x_{n-1}^2 \to x^2, \quad (x_{n-1}^2 + 6) \to (x^2 + 6)$$

and

$$\tfrac{1}{5}(x_{n-1}^2 + 6) \to \tfrac{1}{5}(x^2 + 6)$$

$$\underbrace{\phantom{\tfrac{1}{5}(x_{n-1}^2 + 6)}}_{= x_n}$$

But we know that $x_n \to x$ and so

$$x = \tfrac{1}{5}(x^2 + 6) \quad \text{and} \quad x^2 - 5x + 6 = 0$$

It follows that x is either 2 or 3, but since the sequence starts at 2.5 and decreases it follows that the limit x must be 2.

(Note that we carefully justified the fact that $\tfrac{1}{5}(x_{n-1}^2 + 6)$ converges to $\tfrac{1}{5}(x^2 + 6)$: as we get more confident about which steps are justified we'll be able to make bolder statements like

'letting $n \to \infty$ in $x_n = \tfrac{1}{5}(x_{n-1}^2 + 6)$ gives $x = \tfrac{1}{5}(x^2 + 6)$'.)   □

Example   Let the sequence (x_n) be defined by

$$x_1 = 1 \quad \text{and} \quad x_n = \frac{x_{n-1}}{2} + \frac{1}{[x_{n-1}]} \quad \text{for } n > 1$$

where $[x_{n-1}]$ denotes the 'integer part' of x_{n-1}. Show that

  (i) $1 \leqslant x_n < 2$ for each n;
  (ii) the sequence is increasing;
  (iii) the sequence is convergent, and find its limit.

Solution   If you calculate the first few terms of the sequence you will get $1, 1\tfrac{1}{2}, 1\tfrac{3}{4}, 1\tfrac{7}{8}, 1\tfrac{15}{16}, \ldots$ which is clearly converging to 2. But in order to learn an important lesson we shall proceed exactly as in the previous solution.

(i) We prove that $1 \leqslant x_n < 2$ by induction on n, the case $n = 1$ being trivial. So assume that $1 \leqslant x_{k-1} < 2$. Then $[x_{k-1}] = 1$ and

$$1 \leqslant \tfrac{1}{2} + 1 \leqslant \frac{x_{k-1}}{2} + 1 = \underbrace{\frac{x_{k-1}}{2} + \frac{1}{[x_{k-1}]}}_{= x_k} < \frac{2}{2} + 1 = 2$$

and hence $1 \leqslant x_k < 2$ and the inequalities are established by induction.

(ii) Note that by (i) $1 \leqslant x_{n-1} < 2$ and so $[x_{n-1}] = 1$. Now to show that (x_n) is increasing we'll show that $x_n - x_{n-1}$ is positive:

$$x_n - x_{n-1} = \left(\frac{x_{n-1}}{2} + \frac{1}{[x_{n-1}]} \right) - x_{n-1} = \left(\frac{x_{n-1}}{2} + 1 \right) - x_{n-1}$$

$$= 1 - \frac{x_{n-1}}{2} > 0$$

and so $x_n > x_{n-1}$ and the sequence is increasing.

(iii) We have shown that the sequence (x_n) is increasing and bounded above (by the number 2 for example). Hence the sequence converges with limit x say. Letting '$n \to \infty$ in the equation' we get

$$x_n = \frac{x_{n-1}}{2} + \frac{1}{[x_{n-1}]} \Rightarrow x = \frac{x}{2} + \frac{1}{[x]} \quad \text{or} \quad \frac{x}{2} = \frac{1}{[x]}$$

But the equation $x/2 = 1/[x]$ has no solution: in particular the expected answer of $x = 2$ does not satisfy that equation. Why not? Where did we go wrong? We made the following deduction

$$x_1, x_2, x_3, \ldots \to x \Rightarrow$$

$$\frac{x_1}{2} + \frac{1}{[x_1]}, \quad \frac{x_2}{2} + \frac{1}{[x_2]}, \quad \frac{x_3}{2} + \frac{1}{[x_3]}, \quad \ldots \to \frac{x}{2} + \frac{1}{[x]}$$

but the sequence is $1, 1\frac{1}{2}, 1\frac{3}{4}, 1\frac{7}{8}, 1\frac{15}{16}, \ldots$ and so

$$x_1, x_2, x_3, \ldots \to x$$
$$\| \quad \| \quad \| \qquad \|$$
$$1 \quad 1\tfrac{1}{2} \quad 1\tfrac{3}{4} \qquad 2$$

but

$$\frac{x_1}{2} + \frac{1}{[x_1]}, \quad \frac{x_2}{2} + \frac{1}{[x_2]}, \quad \frac{x_3}{2} + \frac{1}{[x_3]} \ldots \nrightarrow \frac{x}{2} + \frac{1}{[x]}$$
$$\| \qquad\qquad \| \qquad\qquad \| \qquad\qquad \|$$
$$1\tfrac{1}{2} \qquad\qquad 1\tfrac{3}{4} \qquad\qquad 1\tfrac{7}{8} \qquad\qquad 1\tfrac{1}{2}$$

(In the previous example $x_n = f(x_{n-1})$ for some function f and by quoting some earlier results we were able to deduce that

$$x_1, x_2, x_3, \ldots \to x \Rightarrow f(x_1), f(x_2), f(x_3), \ldots \to f(x)$$

But in this example such a deduction was invalid. Certain functions can be 'applied right through a convergent sequence' and others cannot. This is an important topic which we shall return to in the next chapter.)   □

Exercises

1   Are the following sequences convergent? If so find their limits, justifying your answers.

(i) $\left(\dfrac{2n^3 + 1}{3n^3 + n + 2}\right)$ (ii) $\left(\left(1 + \dfrac{1}{\sqrt{n}}\right)^2\right)$ (iii) $((100 + 5^n)^{1/n})$

2   Let the sequence (x_n) be given by

$$x_1 = 10 \quad \text{and} \quad x_n = \frac{x_{n-1}}{2} + \frac{1}{x_{n-1}} \quad \text{for} \quad n > 1$$

We observed in exercise 3 on page 29 that this sequence is decreasing and bounded below. Use those facts to show that the sequence is convergent and hence find its limit.

3   We saw in an earlier example that for $x > 0$ the sequence $(x^{1/n})$ converges to 1. We now wish to investigate the behaviour of the sequence $(n^{1/n})$ or $1, 2^{1/2}, 3^{1/3}, 4^{1/4}, \ldots$. Before proceeding calculate a few terms of the sequence and guess what its limit is going to be.

In exercise (6)(i) on page 19 we established Bernoulli's inequality, namely $1 + nx \leqslant (1 + x)^n$ for each positive integer n and each number $x \geqslant -1$. Use this inequality with $x = 1/\sqrt{n}$ to show that

$$1 \leqslant n^{1/n} < (1 + \sqrt{n})^{2/n} \leqslant (1 + 1/\sqrt{n})^2$$

Deduce that the sequence $(n^{1/n})$ converges to 1.

4   Let x be any positive number and let N be any integer larger than x. Show that from the Nth term onwards the sequence $(x^n/n!)$ is decreasing. Deduce that the sequence is convergent and find its limit.

(Solutions on page 241)

Infinite sums

We commented very early on that our rules of primary arithmetic allow us to work out the sum of three or more numbers without the need for brackets. So, for example, the meaning of

$$1 + \tfrac{1}{2} + \tfrac{1}{4} + \tfrac{1}{8} + \tfrac{1}{16}$$

is unambiguous. If you work out that sum the chances are that you work from left to right calculating subtotals as you go:

$$1, 1\tfrac{1}{2}, 1\tfrac{3}{4}, 1\tfrac{7}{8}, 1\tfrac{15}{16}$$

giving the answer $1\tfrac{15}{16}$. But what if I gave you the sum of an infinite number of terms, such as

$$1 + \tfrac{1}{2} + \tfrac{1}{4} + \tfrac{1}{8} + \tfrac{1}{16} + \cdots?$$

Wouldn't the general principle be the same – you would work from the left, calculating subtotals as you went, giving:

$$1, 1\tfrac{1}{2}, 1\tfrac{3}{4}, 1\tfrac{7}{8}, 1\tfrac{15}{16}, 1\tfrac{31}{32}, 1\tfrac{63}{64}, 1\tfrac{127}{128}, \ldots$$

But then instead of carrying on until you 'reached' an answer you might be tempted to say that these subtotals 'seem to be approaching' 2, and therefore you would guess that the sum of that endless list 'equalled' 2.

None of that is surprising or new: it is merely outlining a procedure you will already have used before. That procedure now enables us to make the following natural definition of the convergence of an infinite sum or 'series':

> **Definition**   To calculate the sum of the *series*
>
> $$a_1 + a_2 + a_3 + \cdots \left(\text{or} \sum_{n=1}^{\infty} a_n \right)$$
>
> calculate the *partial sums* (or subtotals)
>
> $$s_1 = a_1$$
> $$s_2 = a_1 + a_2$$
> $$s_3 = a_1 + a_2 + a_3$$
> $$\vdots$$
> $$s_n = a_1 + a_2 + \cdots + a_n$$
> $$\vdots$$
>
> Then if the sequence $s_1, s_2, s_3, \ldots$ converges (with limit s say) we say that the series $\sum a_n$ *converges* (with *sum s*). If the sequence (s_n) diverges then we say that the series $\sum a_n$ *diverges*.

Important note: If the terms of a series are all non-negative then the sequence of its partial sums is increasing: hence such a series will be convergent if and only if all its partial sums are bounded above.

You should already be familiar with the idea of summing a series because we have seen it at work throughout this chapter. For example to test the convergence of the series

$$1 - \tfrac{1}{2} + \tfrac{1}{4} - \tfrac{1}{8} + \cdots \quad \text{or} \quad \sum_{n=1}^{\infty} (-\tfrac{1}{2})^{n-1} \left(\text{or} \sum_{n=0}^{\infty} (-\tfrac{1}{2})^n \right)$$

you need to look at the sequence of partial sums

$$1, \tfrac{1}{2}, \tfrac{3}{4}, \tfrac{5}{8}, \tfrac{11}{16}, \tfrac{21}{32}, \ldots$$

We went through this process in exercise 1 on page 36 and concluded that the sum of the series is $\tfrac{2}{3}$. Similarly in the theorem on page 20 we showed that the series

$$1 + \frac{1}{1!} + \frac{1}{2!} + \frac{1}{3!} + \frac{1}{4!} + \cdots \quad \text{or} \quad \sum_{n=0}^{\infty} \frac{1}{n!}$$

had a sum of approximately 2.718282. And in exercise 3 on page 22 we saw that the series

$$1 + \tfrac{1}{2} + \tfrac{1}{3} + \tfrac{1}{4} + \cdots \quad \text{or} \quad \sum_{n=1}^{\infty} \frac{1}{n}$$

is not bounded above, or that it tends to infinity, and so that series diverges.

We note that there are some simple properties of series (=plural of series!) which follow trivially from the corresponding results about sequences, and we shall accept these without proof:

(i) if the series $\sum a_n$ converges with sum s and b is any real number, then the series $\sum b a_n$ converges with sum bs (for if the partial sums of $\sum a_n$ are $s_1, s_2, s_3, \ldots$ then the partial sums of $\sum b a_n$ are $b s_1, b s_2, b s_3, \ldots$ and we are back in the familiar world of sequences);

(ii) if the series $\sum a_n$ converges with sum s and the series $\sum b_n$ converges with sum t then the series $\sum (a_n + b_n)$ converges with sum $s + t$, with a similar result for the difference of two series;

(iii) the convergence of a series depends upon the behaviour of its terms at the 'tail-end' and so the first 100 (or 1000 or 1 000 000) terms of the series can be changed without affecting its convergence. Hence the series $\sum_{n=1}^{\infty} a_n$ is convergent if and only if the series $\sum_{n=N}^{\infty} a_n$ is convergent (see the next exercises).

Our next result about the convergence of series does *not* enable you to deduce that any particular series is convergent, but it does enable you to see that certain series are divergent: it states the obvious fact that if the terms in an infinite list do not tend towards zero then there is no hope of adding them all up.

> ***Theorem***   If the sequence $a_1, a_2, a_3, \ldots$ does not converge to 0 then the series $\sum a_n$ diverges.

> ***Pre-proof (a logical aside)***   Consider first the logical statement
>
> **if** Fred is reading this book **then** Fred can read.

That is logically equivalent to the statement

> **if** Fred cannot read **then** Fred is not reading this book.

But it certainly does *not* say

> **if** Fred can read **then** Fred is reading this book

(such a statement would do wonders for its sales!).

So note carefully that this theorem says

> **if** the individual terms don't tend to 0 **then** the series doesn't converge

or

> **if** the series converges **then** the individual terms tend to 0

But it does *not* say

if the individual terms tend to 0 **then** the series converges.

A counter-example to the last statement can be found by considering the series $\sum 1/n$ whose terms tend to 0 but which we know diverges. Having made that important logical point we can now proceed to prove the theorem.

Proof   Assume that the series $\sum a_n$ converges (we shall deduce that its terms $a_1, a_2, a_3, \ldots$ tend to 0). Let s_n be the nth partial sum of the series. Then since the series converges (with sum s say) we know that $s_n \to s$ as $n \to \infty$. But then s_{n-1} also tends to s as $n \to \infty$. Hence

$$a_n = (a_1 + a_2 + \cdots + a_n) - (a_1 + a_2 + \cdots + a_{n-1})$$

$$= s_n - s_{n-1} \to s - s = 0$$

as required.   □

Example   Show that the series $\sum \cos(1/n)$ diverges.

Solution   The individual terms $\cos(1/n)$ tend to 1 (and hence not to 0) as $n \to \infty$. Therefore by the theorem the series diverges.   □

We have seen (exercise 5 on page 19) that

$$\frac{1}{1 \times 2} + \frac{1}{2 \times 3} + \frac{1}{3 \times 4} + \cdots + \frac{1}{n(n+1)}$$

$$= 1 - \frac{1}{n+1} \quad (\to 1 \text{ as } n \to \infty)$$

Hence the series

$$\frac{1}{1 \times 2} + \frac{1}{2 \times 3} + \frac{1}{3 \times 4} + \cdots$$

converges (with sum 1). We then deduced (exercise 2 on page 22) that the 'smaller' series

$$\frac{1}{2 \times 2} + \frac{1}{3 \times 3} + \frac{1}{4 \times 4} + \cdots$$

converges. Hence we have seen that $\sum 1/n^2$ converges by comparing it with a larger convergent series. In general we have the following result:

Theorem   (*The comparison test*) Let $0 \leqslant a_n \leqslant b_n$ for each positive integer n. Then if the series $\sum b_n$ converges it follows that the series $\sum a_n$ also converges.

(Note that by our comments about the 'tail-end' behaviour of series we could weaken the requirement in the theorem to '$0 \leqslant a_n \leqslant b_n$ from some Nth term onwards'.)

Proof   We know from the definition of convergence of the series $\sum b_n$ that the sequence of its partial sums $t_1, t_2, t_3, \ldots$ converges to some number t. Also, since all the terms b_n are non-negative, it follows that the sequence of partial sums $t_1, t_2, t_3 \ldots$ is increasing towards t. Now if $s_1, s_2, s_3, \ldots$ is the sequence of partial sums of the series $\sum a_n$ then

$$s_1 = a_1 \leqslant b_1 = t_1 \leqslant t$$
$$s_2 = a_1 + a_2 \leqslant b_1 + b_2 = t_2 \leqslant t$$
$$s_3 = a_1 + a_2 + a_3 \leqslant b_1 + b_2 + b_3 = t_3 \leqslant t$$
$$\vdots$$

and we see that the increasing sequence $s_1, s_2, s_3, \ldots$ of partial sums of the series $\sum a_n$ is bounded above by t. Hence the sequence (s_n) is convergent (with limit less than or equal to t) which shows that the series $\sum a_n$ is convergent (with sum less than or equal to t). $\square$

Example   Show that for $-1 < r < 1$ the 'geometric' series $\sum_{n=0}^{\infty} r^n$ converges with sum $1/(1-r)$. Deduce that the series $\sum_{n=0}^{\infty} \sin^2 n \sin^n 2$ converges.

Solution   The nth partial sum of the series $\sum_{n=0}^{\infty} r^n$ is given by

$$1 + r + r^2 + r^3 + \cdots r^{n-1} = \frac{1 - r^n}{1 - r} \quad (r \neq 1)$$

But for $-1 < r < 1$ we know that r^n converges to 0 as $n \to \infty$ and so the sequence of partial sums converges to $1/(1-r)$ as required.
    Now

$$0 \leqslant \sin^2 n \sin^n 2 \leqslant \sin^n 2 = (\sin 2)^n$$

and $\sin 2$ is less than 1. Hence by the first part of this example $\sum (\sin 2)^n$ converges and by the comparison test so does the series $\sum \sin^2 n \sin^n 2$. (All that we know about its sum is that it is less than $1/(1 - \sin 2)$.) $\square$

Our next theorem and corollary give another test for convergence of a series of positive terms, known as the *ratio test*. When testing the convergence of the series $\sum a_n$, instead of having to find another series $\sum b_n$ with which to compare it, the ratio test is applied directly to the terms $a_1, a_2, a_3, \ldots$. We have to calculate the 'ratios' a_{n+1}/a_n and hope that they become small. If they are all eventually less than or equal to r, where r is some number less than 1, then the individual terms are getting dramatically smaller and smaller and the series turns out to be convergent. If the ratios are all eventually bigger than or equal to 1, then the terms get out of hand and we'll see that the series diverges. (The in-between case where the ratios are less than 1 but get arbitrarily close to it cannot be decided by the ratio test. For example:

convergent series $\sum 1/n^2$ ratio of $(n + 1)$th term / nth term	**divergent series** $\sum 1/n$ ratio of $(n + 1)$th term / nth term
$= \dfrac{1}{(n + 1)^2} \div \dfrac{1}{n^2} = \dfrac{n^2}{(n + 1)^2} < 1$	$= \dfrac{1}{n + 1} \div \dfrac{1}{n} = \dfrac{n}{n + 1} < 1$
(but arbitrarily close to 1)	(but arbitrarily close to 1)

and we have two series whose ratios behave the same way but one of which converges and one diverges.)

> **Theorem** Let $\sum a_n$ be a series of positive terms and for each n consider the nth ratio a_{n+1}/a_n. Then
>
> (i) if all the ratios are less than or equal to some number r where $r < 1$ then the series converges;
>
> (ii) if all the ratios are bigger than or equal to 1 then the series diverges.
>
> (Again these conditions can be weakened: the ratios need only have their property from some Nth term onwards.)

> **Proof** (i) We have that

$$\frac{a_{n+1}}{a_n} \leqslant r < 1 \quad \text{for each positive integer } n$$

Therefore

$$\frac{a_2}{a_1} \leqslant r, \ \frac{a_3}{a_2} \leqslant r, \ \frac{a_4}{a_3} \leqslant r, \ \ldots$$

and so

$$a_2 \leqslant ra_1, \ a_3 \leqslant ra_2 \leqslant r^2 a_1, \ a_4 \leqslant ra_3 \leqslant r^3 a_1, \ \ldots$$

Now the series $\sum r^{n-1}$ converges and hence so does the series $\sum r^{n-1} a_1$. But

$$0 \leqslant a_n \leqslant r^{n-1} a_1 \quad \text{for each } n$$

Hence by the comparison test the series $\sum a_n$ also converges.

(ii) Here we have

$$\frac{a_2}{a_1} \geqslant 1, \ \frac{a_3}{a_2} \geqslant 1, \ \frac{a_5}{a_4} \geqslant 1, \ \dots$$

But then

$$0 < a_1 \leqslant a_2 \leqslant a_3 \leqslant \cdots$$

and the sequence (a_n) does not converge to 0. Hence by the theorem on page 65 the series $\sum a_n$ diverges.   □

Corollary   (*The ratio test*) Let $\sum a_n$ be a series of positive terms for which the sequence (a_{n+1}/a_n) converges to some limit x. Then:

   (i) if $x < 1$ the series converges;
   (ii) if $x > 1$ the series diverges;

(and if $x = 1$ no conclusion can be drawn either way).

Proof   (i) If a sequence converges to $x < 1$ then $\frac{1}{2}(1 - x) > 0$ and so taking $\varepsilon = \frac{1}{2}(1 - x)$ in the definition of convergence shows that from some Nth term onwards all the terms of the sequence will lie between $x - \frac{1}{2}(1 - x)$ and $x + \frac{1}{2}(1 - x)$ (which equals $\frac{1}{2}(1 + x)$). In particular those terms will be less than $\frac{1}{2}(1 + x)$ which is less than 1.

Hence the terms of a_{n+1}/a_n are all eventually less than $\frac{1}{2}(1 + x)$ which is itself less than 1, and we can apply the theorem to show that the series $\sum a_n$ converges.

(ii) Similarly if a sequence converges to $x > 1$ then from some Nth term onwards all its terms will be greater than 1. Hence the terms of a_{n+1}/a_n are all eventually greater than 1 and by the theorem the series a_n diverges.   □

Example   Examine the convergence of the series

(i) $\sum \dfrac{(2n)!}{(n!)^2}$;   (ii) $\sum \dfrac{(3n)!}{7^n(2n)!n!}$

Solution   (i) In this case the ratio of the $(n + 1)$th term to the nth is

$$\frac{(2(n + 1))!}{((n + 1)!)^2} \div \frac{(2n)!}{(n!)^2} = \frac{(2n + 2)(2n + 1)}{(n + 1)(n + 1)} = 2\frac{2 + 1/n}{1 + 1/n} \to 4 \text{ as } n \to \infty$$

As the limit exists and is greater than 1 the corollary tells us that the series diverges.

(ii) In this case the ratio of successive terms is

$$\frac{(3(n+1))!}{7^{n+1}(2(n+1))!(n+1)!} \div \frac{(3n)!}{7^n(2n)!n!} = \frac{(3n+3)(3n+2)(3n+1)}{7(2n+2)(2n+1)(n+1)}$$

$$= \frac{3(3+2/n)(3+1/n)}{7(2+2/n)(2+1/n)} \to \frac{27}{28} \quad \text{as } n \to \infty$$

Since the limit is less than 1 the corollary shows that the series converges.

□

If a series of non-negative terms converges then, as you might guess, changing the signs of some of the terms to make some of them negative will, if anything, make the new series more likely to 'add up' or converge:

> **Theorem**   If $a_1, a_2, a_3, \ldots$ are numbers such that the series $\sum |a_n|$ converges, then the series $\sum a_n$ also converges: ($\sum a_n$ is the same series as $\sum|a_n|$ except that some of the terms may have been made negative).

Proof   Write down the series $a_1 + a_2 + a_3 + \cdots$ but put the non-negative terms in one row and the negative terms in another. For example the series $1 + \frac{1}{4} - \frac{1}{9} + \frac{1}{16} + \frac{1}{25} - \frac{1}{36} + \frac{1}{49} + \cdots$ would be written

$$1 + \tfrac{1}{4} \qquad + \tfrac{1}{16} + \tfrac{1}{25} \qquad + \tfrac{1}{49} + \tfrac{1}{64} \qquad \cdots$$
$$\qquad -\tfrac{1}{9} \qquad\qquad -\tfrac{1}{36} \qquad\qquad -\tfrac{1}{81} \cdots$$

Fill in the gaps in each row with 0s and make all the minus signs in the second row into plus signs, giving two new series $\sum p_n$ and $\sum q_n$. In the above example these would be

$$\sum p_n = 1 + \tfrac{1}{4} + 0 + \tfrac{1}{16} + \tfrac{1}{25} + 0 \ + \tfrac{1}{49} + \tfrac{1}{64} + \ 0 + \cdots$$
$$\sum q_n = 0 + 0 + \tfrac{1}{9} + 0 \ + 0 \ + \tfrac{1}{36} + 0 \ + 0 \ + \tfrac{1}{81} + \cdots$$

(Note that in each case $p_n - q_n$ equals the original a_n.)

Now for each n

$$0 \leqslant p_n \leqslant |a_n| \quad \text{and} \quad 0 \leqslant q_n \leqslant |a_n|$$

equals 0 if $a_n \leqslant 0$ equals 0 if $a_n \geqslant 0$
or a_n if $a_n > 0$ or $-a_n$ if $a_n < 0$

and so, as we are given that the series $\sum |a_n|$ converges, we can deduce from the comparison test that the series $\sum p_n$ and $\sum q_n$ both converge. But then the series $\sum (p_n - q_n)$ also converges. But if you think about the way we constructed p_n and q_n you will see that $p_n - q_n$ is just a_n. We have therefore deduced that $\sum a_n$ converges as required.

□

A series $\sum a_n$ for which the sum of the 'absolute values' – or moduli – $|a_1| + |a_2| + |a_3| + \cdots$ converges is called *absolutely convergent*. So our comment that you can throw some minus signs into a convergent series of non-negative terms without upsetting its convergence has become a rather posh theorem 'an absolutely convergent series is convergent'.

> **Example**   Show that for any number x the series $\sum x^n/n!$ converges.

> **Solution**   We shall show that this series is absolutely convergent; i.e. that the series $\sum |x^n/n!|$ is convergent. From the above theorem it will then follow that the series $\sum x^n/n!$ is convergent as required.

Now if x is 0 the convergence of the series is immediate. So assume that x is not 0. Then the ratio of successive terms of the series $\sum |x^n/n!|$ is given by

$$\left| \frac{x^{n+1}}{(n+1)!} \right| \div \left| \frac{x^n}{n!} \right| = \frac{|x|}{n+1} \to 0 \quad \text{as } n \to \infty$$

Hence by the ratio test the series $\sum |x^n/n!|$ converges; i.e. the series $\sum x^n/n!$ is absolutely convergent and hence convergent. $\square$

Our final theorem about series concerns ones in which the terms alternate in sign, such as

$$1 - \tfrac{1}{2} + \tfrac{1}{3} - \tfrac{1}{4} + \tfrac{1}{5} - \cdots$$

> **Theorem**   (*The alternating series test*)   Let (a_n) be a decreasing sequence which converges to 0. Then the series
>
> $$a_1 - a_2 + a_3 - a_4 + \cdots \quad (\text{or } \sum (-1)^{n-1} a_n)$$
>
> is convergent.

> **Proof**   Let s_n be the nth partial sum of the given series. Then, since the sequence (a_n) is decreasing,

$$s_{2n+2} - s_{2n} = (a_1 - a_2 + \cdots + a_{2n+1} - a_{2n+2}) - (a_1 - a_2 + \cdots + a_{2n})$$

$$= a_{2n+1} - a_{2n+2} \geqslant 0$$

Hence $s_{2n+2} \geqslant s_{2n}$ for each n and so the sequence $s_2, s_4, s_6,$ is increasing. Also

$$s_{2n} = a_1 - \underbrace{(a_2 - a_3)}_{\geqslant 0} - \underbrace{(a_4 - a_5)}_{\geqslant 0} - \cdots \leqslant a_1.$$

Therefore the sequence (s_{2n}) is also bounded (by a_1) and so the sequence (s_{2n}) converges, to s say.

It then follows that

$$s_{2n+1} = a_1 - a_2 + \cdots - a_{2n} + a_{2n+1} = s_{2n} + a_{2n+1} \to s + 0 = s$$

and so the sequence (s_{2n+1}) also converges to s.

We saw in exercise 2 on page 50 that when the two sequences (s_{2n}) and (s_{2n+1}) converge to the same limit it follows that the sequence (s_n) is convergent. Hence the series $\sum a_n$ converges as required. $\square$

Note that if a series of positive terms converges then changing any number of the positive signs to negative signs will not upset the convergence: this is the result 'absolute convergence implies convergence' proved above. In fact if the convergent series of positive terms has sum s then changing *all* its signs would give a sum of $-s$, and changing just some of the signs would give sums between $-s$ and s.

We have also seen in that last example that although the series

$$1 + \tfrac{1}{2} + \tfrac{1}{3} + \tfrac{1}{4} + \tfrac{1}{5} + \cdots$$

diverges, by changing half the signs we get a convergent series

$$1 - \tfrac{1}{2} + \tfrac{1}{3} - \tfrac{1}{4} + \tfrac{1}{5} - \cdots$$

This latter series is therefore convergent but not absolutely convergent: such a series is called *conditionally convergent*. For such series there is a very fine balance between convergence and divergence and, as we shall see in the exercises, they have some rather odd properties.

We have at last covered the basic theory of sequences to enable us to continue to a full study of functions. We have seen various ways in which sequences are going to be a crucial tool in that study. In addition we can use sequences and series actually to *define* functions. For example we have seen that for any number x the series $\sum_{n=0}^{\infty} x^n/n!$ (or $1 + x/1! + x^2/2! + x^3/3! + \cdots$) converges and so has a sensible sum: we could define a function, 'exp' say, by

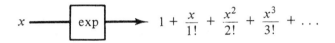

You may have seen this function before: $\exp(1)$ is the number e we constructed in chapter 1. The role of e in this function, and the remarkable connection between the functions exp and log are some of the many fascinating aspects of functions which we are now ready to study.

Exercises

1   Test the convergence of the following series:

(i) $\sum \dfrac{(n^2 + 1)^3}{(n^4 + 1)^2}$;   (ii) $\sum \dfrac{5^{2n}(n!)^3}{(3n)!}$;   (iii) $\sum \sin n$

2   Let $\sum a_n$ be a convergent series with sum s. Let $\sum b_n$ be another series which is the same as the series $\sum a_n$ from the Nth term onwards. Let

$$t = (b_1 + b_2 + \cdots + b_{N-1}) - (a_1 + a_2 + \cdots + a_{N-1})$$

Show that the series $\sum b_n$ converges with sum $s + t$.

3   Show that the series $\sum 1/n^r$ diverges for $r \leqslant 1$ and converges for $r > 1$.

(For the latter part you will need the result of exercise 3 on page 22.)

4   Let $\sum a_n$ be a series with partial sums $s_1, s_2, s_3, \ldots$. Show that if the sequence $s_2, s_4, s_6, \ldots$ converges and the terms $a_1, a_2, a_3, \ldots$ converge to 0 then the series $\sum a_n$ converges. (If you're keen show that the condition that (s_{2n}) converges can be reduced to the fact that (s_{Mn}) converges for some fixed positive integer M.)

5   We know from our primary arithmetic that changing the order of a sum of a finite number of terms will not affect the answer; e.g.

$$7 - 3\tfrac{1}{2} + \tfrac{1}{4} - 1 - 2 = \tfrac{3}{4} = -1 - 3\tfrac{1}{2} + 7 - 2 + \tfrac{1}{4}$$

and the same actually turns out to be true of any rearrangement of an absolutely convergent series. But for those series which are only conditionally convergent rearranging the order of the terms in them can actually change their sum! This exercise illustrates that fact.

Let s_n be the nth partial sum of the series

$$1 - \tfrac{1}{2} + \tfrac{1}{3} - \tfrac{1}{4} + \tfrac{1}{5} - \tfrac{1}{6} + \tfrac{1}{7} - \tfrac{1}{8} + \cdots$$

(We know that this series converges, with sum s say.)

Let t_n be the nth partial sum of the rearranged series

$$1 - \tfrac{1}{2} - \tfrac{1}{4} + \tfrac{1}{3} - \tfrac{1}{6} - \tfrac{1}{8} + \tfrac{1}{5} - \tfrac{1}{10} - \cdots$$

Show that $t_{3n} = \tfrac{1}{2}s_{2n}$.

Use exercise 4 to deduce that this latter series converges with sum $\tfrac{1}{2}s$; i.e. by rearranging the series we have halved its sum!

6   We have defined the function log (on page 47) but have not yet established that it has any of the expected logarithmic-type properties. If you're happy to accept for the moment that $\log 2n - \log n$ is $\log(2n/n)$ which is log 2, then you are actually able to calculate the sums of the series in exercise 5.

As before let s_n be the nth partial sum of the series

$$1 - \tfrac{1}{2} + \tfrac{1}{3} - \tfrac{1}{4} + \tfrac{1}{5} - \tfrac{1}{6} + \tfrac{1}{7} - \tfrac{1}{8} + \cdots$$

and also let u_n be the nth partial sum of the series

$$1 + \tfrac{1}{2} + \tfrac{1}{3} + \tfrac{1}{4} + \tfrac{1}{5} + \tfrac{1}{6} + \tfrac{1}{7} + \tfrac{1}{8} + \cdots$$

(i) Use the result of the example on pages 46 and 47 to show that the sequence $(u_n - \log n)$ converges. (Its limit is again γ, Euler's constant.)

(ii) Show that

$$s_{2n} = u_{2n} - u_n = (u_{2n} - \log(2n)) - (u_n - \log n) + \log 2$$

(iii) Deduce that the sum of the series

$$1 - \tfrac{1}{2} + \tfrac{1}{3} - \tfrac{1}{4} + \tfrac{1}{5} - \tfrac{1}{6} + \tfrac{1}{7} + \tfrac{1}{8} + \cdots$$

is $\log 2$ and (from exercise 5) that the sum of the series

$$1 - \tfrac{1}{2} - \tfrac{1}{4} + \tfrac{1}{3} - \tfrac{1}{6} - \tfrac{1}{8} + \tfrac{1}{5} - \tfrac{1}{10} - \cdots$$

is $\tfrac{1}{2} \log 2$. Use your calculator or computer to work out some of the partial sums of these series to confirm directly that these answers look reasonable.

7   For which numbers x does the series $\sum_{n=0}^{\infty} x^n$ converge? If D is the set of x for which that series converges then we can define a function f by

$$f(x) = \sum_{n=0}^{\infty} x^n = 1 + x + x^2 + x^3 + \cdots \quad \text{for } x \in D$$

Give a more direct formulation of f which does not involve infinite sums.

(Solutions on page 243)

3

A functional approach

A powerful start

Later in this chapter we shall discuss the general behaviour of functions but it is more natural to begin by considering some specific examples. We have already encountered the function given by

$$f(x) = x^2 \quad x \in \mathbb{R}$$

The function has domain $\mathbb{R}$ (i.e. it will accept any $x \in \mathbb{R}$) and the set of answers eventually output by f equals the set $[0, \infty)$ of non-negative numbers. This set is called the '*range*' (or '*image*') of f.

Now for any positive integer n we can define a function by

$$f(x) = x \times x \times \cdots \times x = x^n \quad x \in \mathbb{R}$$
$$\longleftarrow n \text{ times} \longrightarrow$$

The graphs of a few such functions are illustrated below.

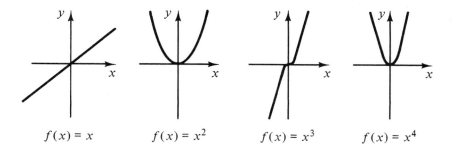

$$f(x) = x \qquad f(x) = x^2 \qquad f(x) = x^3 \qquad f(x) = x^4$$

To extend that idea to negative indices, recall from the rules concerning powers that $x^{-n} = 1/x^n$. Therefore, again for any positive integer n, we can

define a function by

$$f(x) = x^{-n} = \frac{1}{x^n} \quad x \in \mathbb{R}, x \neq 0$$

and the graphs of a few such functions are shown below.

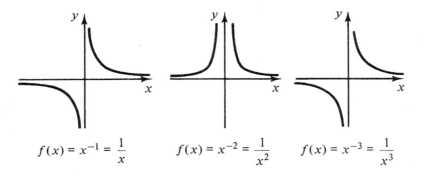

$$f(x) = x^{-1} = \frac{1}{x} \qquad f(x) = x^{-2} = \frac{1}{x^2} \qquad f(x) = x^{-3} = \frac{1}{x^3}$$

So, with the additional understanding that x^0 is 1, for any integer n we can define a function by the rule $f(x) = x^n$.

But what about expressions like $x^{1.5}$ and $x^{1.6}$? They mean

$$x^{1.5} = x^{3/2} = (x^{1/2})^3 = (\sqrt{x})^3 \quad \text{and} \quad x^{1.6} = x^{8/5} = (x^{1/5})^8 = (\sqrt[5]{x})^8$$

and so in order to be able to define $x^{m/n}$ for any rational number m/n we need first to understand the meaning of $x^{1/n}$.

By $x^{1/2}$ we mean the 'square root' of x. We shall only attempt that operation for non-negative x and we shall always choose the non-negative root as our answer. So we can define a function g by

$$g(x) = x^{1/2} \quad x \geqslant 0$$

and observe its graph:

If we now restrict the domain of the function given by $f(x) = x^2$ to the non-negative x then we can investigate the precise relationship between the two functions

$$f(x) = x^2 \quad x \geqslant 0 \qquad \text{and} \qquad g(x) = x^{1/2} \quad x \geqslant 0$$

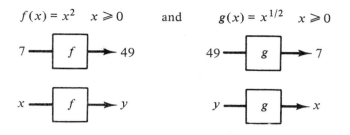

We see that $f(x) = y$ if and only if $g(y) = x$. As you probably know, such functions are 'inverses' of each other. The point (x, y) is on the graph of f if and only if the point (y, x) is on the graph of g. So the graph of g is obtained from that of f by reversing the two axes (which is equivalent to reflecting the graph in the line $y = x$).

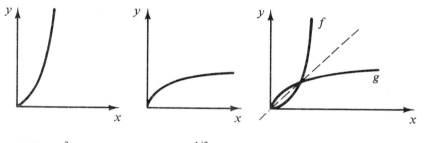

$f(x) = x^2 \quad x \geqslant 0 \qquad g(x) = x^{1/2} \quad x \geqslant 0 \qquad$ graphs of f and g together

In general if a function f has domain D and range G then for each $y \in G$ there exists at least one $x \in D$ with $f(x) = y$. Sometimes (as in the example illustrated on the left below) there may be more than one x with that property. But for some functions, given any $y \in G$ there is a *unique* $x \in D$ with $f(x) = y$: such functions are called *one-to-one* (or *bijections* from D to G), and one such is illustrated on the right below.

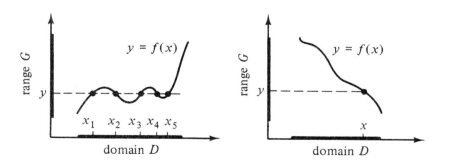

For a one-to-one function with domain D and range G we can define a new function called the *inverse* of f and denoted by f^{-1}. It has domain G and range D and it is defined by the rule

$$f^{-1}(x) = \text{'the unique } y \in D \text{ with } f(y) = x\text{'} \quad x \in G$$

Hence

$$f^{-1}(x) = y \iff f(y) = x$$

and the graphs of f and f^{-1} are mirror-images in the line $y = x$ (as illustrated). Also $f(f^{-1}(x)) = x$ for $x \in G$ and $f^{-1}(f(x)) = x$ for $x \in D$.

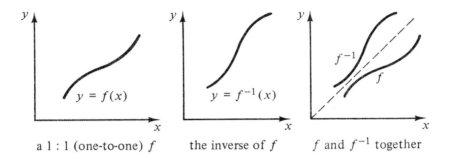

a 1 : 1 (one-to-one) f the inverse of f f and f^{-1} together

We can now explain precisely what we mean by 'the nth root'. Given a positive integer n the function

$$f(x) = x^n \quad x \geq 0$$

has domain and range equal to $[0, \infty)$ and also it is a one-to-one function. (In fact if $x_1 < x_2$ then $f(x_1) < f(x_2)$ and so $f(x)$ never repeats itself.) Hence f has a well-defined inverse function which is denoted by

$$f^{-1}(x) = x^{1/n} \text{ ('the } n\text{th root of } x\text{')} \quad x \geq 0$$

As an example the graphs of the cube and cube root functions are illustrated:

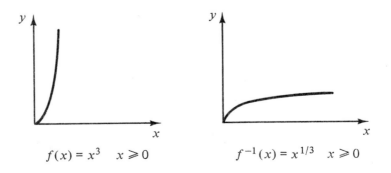

$f(x) = x^3 \quad x \geq 0$ $f^{-1}(x) = x^{1/3} \quad x \geq 0$

Now given any rational number m/n (where m and n are integers and $n > 0$) we are able to define a function by the rule

$$f(x) = x^{m/n} = (x^{1/n})^m \quad x > 0$$

The graphs of the function $f(x) = x^{1.41}$ and $g(x) = x^{1.42}$ are illustrated on the same axes below. For the low values of x shown the graphs are virtually indistinguishable.

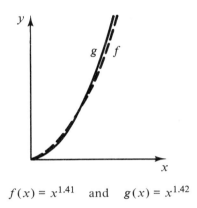

$$f(x) = x^{1.41} \quad \text{and} \quad g(x) = x^{1.42}$$

But what do we mean by $x^{\sqrt{2}}$? We would perhaps expect it to be similar to those two graphs and somewhere between the two of them. We leave this idea of x^s for an irrational number s until the next section.

Exercises

1   Which of these functions is one-to-one? Find the inverse functions in those cases.

(i) $f(x) = x^2 \quad x \in \mathbb{R}$

(ii) $g(x) = 1/x \quad x \in \mathbb{R}, x \neq 0$

(iii) $h(x) = \dfrac{2x - 1}{x + 1} \quad x \in \mathbb{R}, x \neq -1$

That second example is strangely related to its inverse: can you find some other functions with the same property?

2   A function is called *strictly increasing* if whenever x_1 and x_2 are in its domain with $x_1 < x_2$ it follows that $f(x_1) < f(x_2)$. (Similarly it is *strictly decreasing* if $x_1 < x_2$ implies $f(x_1) > f(x_2)$.) Show that a function which is strictly increasing is one-to-one and that its inverse function is

also strictly increasing. Prove also the corresponding result about a decreasing function.

Give an example of a one-to-one function which is neither increasing nor decreasing.

3   (i) Let x_1 and x_2 be any real numbers and let α and β be non-negative numbers with $\alpha + \beta = 1$. Note that for any real number T $(T + x_1)^2 \geqslant 0$ and $(T + x_2)^2 \geqslant 0$ and hence that

$$\alpha(T + x_1)^2 + \beta(T + x_2)^2 \geqslant 0$$

Multiply out that last expression and write it as a quadratic $aT^2 + bT + c$ for some a, b and c. Since that quadratic is never negative use the '$b^2 - 4ac$' to show that

$$(\alpha x_1 + \beta x_2)^2 \leqslant \alpha x_1{}^2 + \beta x_2{}^2$$

(ii) Let f be a function with domain equal to some interval. Then (as we defined in exercise 3 on page 37) f is 'convex' if each of the chords of its graph lies on or above the graph.

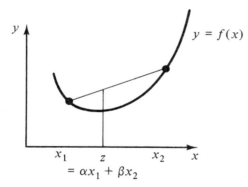

By recalling that each number between x_1 and x_2 is of the form $\alpha x_1 + \beta x_2$ for some non-negative α and β with $\alpha + \beta = 1$, show that f is convex if and only if

$$f(\alpha x_1 + \beta x_2) \leqslant \alpha f(x_1) + \beta f(x_2)$$

for each x_1, x_2 in f's domain and each $\alpha, \beta \geqslant 0$ with $\alpha + \beta = 1$.

(iii) Show that the function g given by $g(x) = x^2$ ($x \in \mathbb{R}$) is convex.

(iv) By using a similar condition to that in (ii) but for a concave function show that the function h given by $h(x) = \sqrt{x}$ ($x \geqslant 0$) is concave.

(You might like to consider more generally the inverse of a strictly increasing convex function and show that it is concave.)

(Solutions on page 245)

Exponentiation

That's just the posh word for the act of raising a number to a power. For the purposes of illustration we shall raise the number 10 to various powers. We recalled in the previous section that $10^{m/n}$ made sense for any rational number m/n. So now we can define a function by

$$f(x) = 10^x \quad x \in \mathbb{Q}$$

(Note that this is quite different from the function $f(x) = x^{10}$: now the variable is the power itself.) Before looking at the graph of f recall that the rational numbers are 'dense' amongst the real numbers in the sense that between any two numbers there is a rational: the irrationals have the same property. Therefore it is easy to see that for any $x \in \mathbb{R}$ there are rationals (and irrationals) arbitrarily close to x. So when we draw the graph of a function with domain $\mathbb{Q}$, the set of rationals, even though it is theoretically riddled with gaps these are impossible to spot with the naked eye.

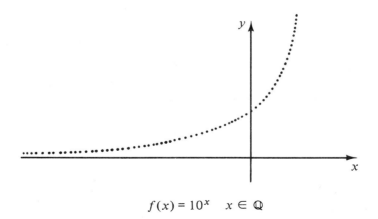

$$f(x) = 10^x \quad x \in \mathbb{Q}$$

We can now begin to study the analytical behaviour of this strictly increasing function.

> ***Example***   (i) Let $\varepsilon > 0$. Show that if M is an integer with $(1 + \varepsilon)^M > 10$ and $\delta = 1/M$ then for any rational number x with $-\delta < x < \delta$ we have $1 - \varepsilon < 10^x < 1 + \varepsilon$.
>
> (ii) Deduce that if $x_1, x_2, x_3, \ldots$ is a sequence of rational numbers which converges to 0 then the sequence 10^{x_1}, 10^{x_2}, $10^{x_3}, \ldots$ converges to 1.

> ***Solution***   (i) Since $\varepsilon > 0$ we have $1 + \varepsilon > 1$ and the sequence $(1 + \varepsilon)$, $(1 + \varepsilon)^2$, $(1 + \varepsilon)^3, \ldots$ tends to infinity. In particular there is a

positive integer M with $(1 + \varepsilon)^M > 10$. Let $\delta = 1/M$ and let x be any rational number between $-\delta$ and δ. Then we have

$$\frac{1}{10^{1/M}} = 10^{-\delta} < 10^x < 10^\delta = 10^{1/M}$$

But since $(1 + \varepsilon)^M > 10$ we then have $10^{1/M} < 1 + \varepsilon$ and so

$$1 - \varepsilon = \frac{1 - \varepsilon^2}{1 + \varepsilon} < \frac{1}{1 + \varepsilon} < \frac{1}{10^{1/M}} < 10^x < 10^{1/M} < 1 + \varepsilon$$

and

$$1 - \varepsilon < 10^x < 1 + \varepsilon$$

as required.

Before proceeding to the second part of the solution let us illustrate that first result.

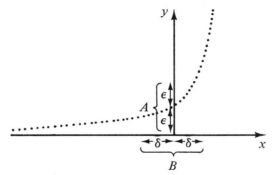

Given an arbitrary $\varepsilon > 0$ we have found a $\delta > 0$ such that if x is in region B then it follows that 10^x will be in region A.

(ii) Now let $x_1, x_2, x_3, \ldots$ be any sequence of rationals which converges to 0. We are asked to show that the sequence $10^{x_1}, 10^{x_2}, 10^{x_3}, \ldots$ converges to 1 and so we take an $\varepsilon > 0$ and try to find an N with

$$1 - \varepsilon < 10^{x_n} < 1 + \varepsilon$$

for all $n \geq N$.

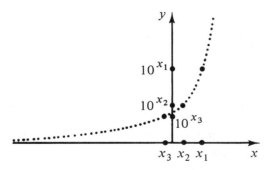

But in (i) we found a positive number δ such that if x is within δ of 0 then 10^x is within ε of 1. Since the sequence (x_n) converges to 0 it follows that there exists an integer N with

$$0 - \delta < x_n < 0 + \delta$$

for all $n \geqslant N$.

But then since all of $x_N, x_{N+1}, x_{N+2}, \ldots$ are within δ of 0 it follows that all of $10^{x_N}, 10^{x_{N+1}}, 10^{x_{N+2}} \ldots$ are within ε of 1. Hence the sequence (10^{x_n}) converges to 1 as required.

So whenever a sequence $x_1, x_2, x_3, \ldots$ converges to 0 it follows that the sequence $10^{x_1}, 10^{x_2}, 10^{x_3}, \ldots$ converges to 1. In the next section we shall meet this idea in general and say that '10^x tends to 1 as x tends to 0' or that '$\lim_{x \to 0} 10^x = 1$'.   □

We are now ready to decide formally what we mean by $10^{\sqrt{2}}$ (and 10^{π} and in general 10^x for any irrational number x). To a non-pure mathematician that would cause no problem at all. Either he/she would look at the graph of $f(x) = 10^x$ $(x \in \mathbb{Q})$, ignore the fact that there is no actual point on the graph where $x = \sqrt{2}$, assume that the graph was in one uninterrupted piece, and read off the value of $10^{\sqrt{2}}$; this is shown on the left below.

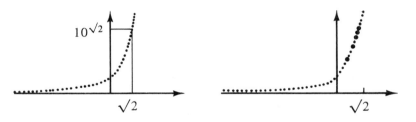

Alternatively he/she would use a calculator to work out $\sqrt{2}$ (ignoring the fact that this only gives a rational approximation to $\sqrt{2}$) and then use that number to obtain the approximate value of $10^{\sqrt{2}}$. To get better and better approximations in this way one would need closer and closer rational approximations to $\sqrt{2}$: e.g.

$$1, \ 1.4, \ 1.41, \ 1.414, \ 1.4142, \ 1.41421, \ 1.414213, \ldots$$

giving

$10^1, \ 10^{1.4}, \quad 10^{1.41}, \quad \ 10^{1.414}, \quad \ 10^{1.4142}, \quad 10^{1.41421}, \ 10^{1.414213}, \ldots$

‖   ‖   ‖   ‖   ‖   ‖   ‖

10   25.118   25.7040   25.94180   25.953743   25.954341   25.95452, …

Intuitively this method is moving along the graph getting closer and closer to the gap at $x = \sqrt{2}$, as shown on the right above.

Both those methods are aimed at finding the 'obvious' value of $10^{\sqrt{2}}$ (and similarly of 10^x for any irrational x) so that when those values are added to the graph of f we still get a strictly increasing function which is just the original graph 'with the gaps filled in'.

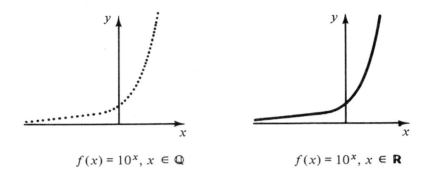

$$f(x) = 10^x, \; x \in \mathbb{Q} \qquad\qquad f(x) = 10^x, \; x \in \mathbb{R}$$

Now as pure mathematicians we are going to be able to say what we mean by $10^{\sqrt{2}}$ by making the above process precise and analytical. As in the second method above we shall let (x_n) be any sequence of rational numbers converging to $\sqrt{2}$ and we would like to *define* $10^{\sqrt{2}}$ as the limit of the sequence (10^{x_n}). But for pure mathematicians like us that process immediately produces warning bells ringing in our ears. There are two crucial questions which have to be answered:

(i) Will the sequence (10^{x_n}) definitely converge?

(ii) If someone else chooses a different sequence (x_n') converging to $\sqrt{2}$ and then takes $10^{\sqrt{2}}$ as the limit of $(10^{x_n'})$ will they get the same answer as us?

> ***Example*** (i) Let (x_n) be an increasing sequence of rational numbers which converges to $\sqrt{2}$. Show that the sequence (10^{x_n}) converges, and call its limit l.
>
> (ii) Let (x_n') be any sequence of rational numbers which converges to $\sqrt{2}$. Show that the sequence $(10^{x_n'})$ also converges to the same number l.

Solution (i) If the sequence (x_n) is increasing to $\sqrt{2}$ then the sequence (10^{x_n}) is also increasing and it is bounded above (by 100 for example). Hence the sequence (10^{x_n}) converges, to l say.

(ii) Now if (x_n') is any sequence of rationals converging to $\sqrt{2}$ then the sequence (x_n') is bounded and hence so is the sequence $(10^{x_n'})$. Also the sequence of rationals $(x_n - x_n')$ converges to 0 $(=\sqrt{2} - \sqrt{2})$ and so by the previous example $(10^{x_n - x_n'})$ converges to 1. We therefore have

$$10^{x_n'} = (10^{x_n'} - 10^{x_n}) + 10^{x_n}$$

$$= \underbrace{10^{x_n'}}_{\text{bounded}} \underbrace{(1 - 10^{x_n - x_n'})}_{\text{convergent to } 0} + \underbrace{10^{x_n}}_{\text{convergent to } l}$$

$\therefore$ convergent to 0

$\therefore$ convergent to l

Hence regardless of the choice of the sequence (x_n') converging to $\sqrt{2}$ the sequence $(10^{x_n'})$ converges to the same limit l and we may take that value unambiguously as the value of $10^{\sqrt{2}}$.   □

We shall explore those ideas further in the next section where the above situation will be described by saying that '10^x tends to l as x tends to $\sqrt{2}$' or '$\lim_{x \to \sqrt{2}} 10^x = l$'.

We know that the rationals are 'dense' in $\mathbb{R}$ and in particular given any number x there is a sequence of rationals converging to x. So by the sort of limit-taking process just described we are able to extend the ideas of 10^x to all real x. In general, as in the examples above,

if $x_1, x_2, x_3, \ldots \to x$ **then** $10^{x_1}, 10^{x_2}, 10^{x_3}, \ldots \to 10^x$

In this way we have defined a function by the rule

$$f(x) = 10^x \quad x \in \mathbb{R}$$

This function has domain $\mathbb{R}$ and range $(0, \infty)$ and, as we shall see in the exercises, it is strictly increasing (which you might have guessed from its graph).

It follows from exercise 2 above that f has a well-defined strictly increasing inverse function with domain $(0, \infty)$ and range $\mathbb{R}$.

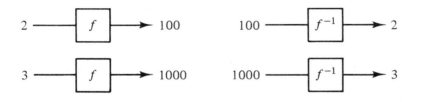

In general for $x > 0$ we have

$$f^{-1}(x) = y \quad \Leftrightarrow \quad x = f(y) = 10^y$$

and so $f^{-1}(x)$ is 'the power to which 10 must be raised in order to get the answer x'. As you will probably already know this is called the *logarithm* of *x to the base 10* and it is denoted by $\log_{10}(x)$ (or simply $\log_{10} x$). Hence

$$y = \log_{10} x \quad \Leftrightarrow \quad x = 10^y$$

For example $\log_{10} 0.01 = -2$, $\log_{10} 1 = 0$, $\log_{10} 10 = 1$, $\log_{10} 100 = 2$, etc. and we can draw the graph of the function f^{-1}, related to the graph of f in the usual way:

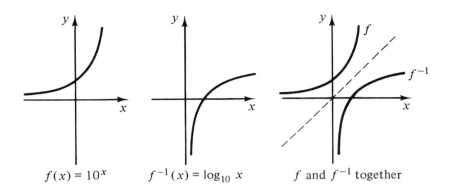

$$f(x) = 10^x \qquad f^{-1}(x) = \log_{10} x \qquad f \text{ and } f^{-1} \text{ together}$$

As we shall see in the next exercises the function $\log_{10}$ has the following multiplicative property:

$$\log_{10} xy = \log_{10} x + \log_{10} y \quad \text{for any positive } x \text{ and } y$$

In the days before calculators this was an indispensible property in the construction of 'slide rules' and 'log tables'. For example imagine that you had a (very large) table of all the values of $\log_{10}$ to (say) three significant figures. Then some of the values in that table would be as shown below:

x	1.99	2.69	3.52	7.39	9.47
$\log_{10} x$	0.299	0.430	0.546	0.869	0.976

Example   Use the extract from log tables shown above to calculate the approximate values of

(i) 2.69×3.52;   (ii) 2.69×7.39

Solution   (i) We know that

$\log_{10}(2.69 \times 3.52) = \log_{10} 2.69 + \log_{10} 3.52$
$\approx 0.430 + 0.546 = 0.976$

and so (using the table in reverse) 2.69×3.52 is approximately 9.47. (That reverse process was known as 'anti-logging' but we now see that is simply saying that $10^{0.976}$ is approximately 9.47.)

(ii) Similarly

$\log_{10}(2.69 \times 7.39) = \log_{10} 2.69 + \log_{10} 7.39$
$\approx 0.430 + 0.869 = 1.299$

Now the number 1.299 does not appear as a log in our table but note that our required answer is

$$10^{1.299} = 10^1 \times 10^{0.299} \approx 10 \times 1.99 = 19.9 \qquad \square$$

That example was included for historical interest only: with the advent of the electronic calculator the log table has been banished to the museum along with the abacus.

In all that we've said above there is nothing special about the number 10. We could have considered the function $f(x) = 2^x$ ($x \in \mathbb{R}$) and its inverse function $\log_2$. (The only place where the 10 made things easier was in that last example on log tables.) In general we could have taken any number $a > 1$ and looked at the pair of inverse functions a^x and $\log_a x$ – giving logarithms to the base a. (Actually to call a^x or $\log_a x$ a function is a lazy use of the language: I mean the functions whose values at x are a^x and $\log_a x$ – but I'm sure you realised that!) The behaviour would have been very similar to the case $a = 10$ which we chose to consider in detail. We could even have chosen a with $0 < a < 1$ but then the two functions would be strictly decreasing rather than increasing. We illustrate the graphs in the cases $a = 3$ and (overleaf) $a = \frac{1}{3}$.

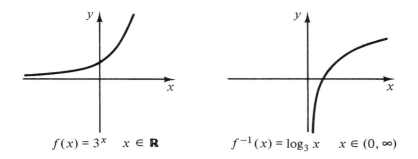

$f(x) = 3^x \quad x \in \mathbb{R}$ $f^{-1}(x) = \log_3 x \quad x \in (0, \infty)$

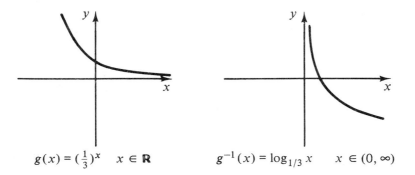

$$g(x) = (\tfrac{1}{3})^x \quad x \in \mathbb{R}$$
$$g^{-1}(x) = \log_{1/3} x \quad x \in (0, \infty)$$

You might be wondering about e^x and $\log_e$: discussion about these will have to wait until we have done a little more work on limits of functions.

Exercises

1   Let $a > 0$. We know (almost from our primary school arithmetic) that

$$a^x \times a^y = a^{x+y} \quad \text{and} \quad a^x \div a^y = a^{x-y}$$

for each pair of rational numbers x and y. By considering sequences of rationals show that these rules extend to *any* pair of numbers x and y. Deduce that if $a > 1$ than the function $f(x) = a^x$ $(x \in \mathbb{R})$ is strictly increasing and that if $0 < a < 1$ the function is strictly decreasing.

2   Let f be any function with domain $\mathbb{R}$ for which:

(i) $f(1)$ is positive (call it a say);
(ii) $f(x) \times f(y) = f(x + y)$ for each pair of rationals x and y.

By first dealing with the integers show that $f(x) = a^x$ for each rational number x.

Now assume in addition that f has the property that

if $x_1, x_2, x_3, \ldots \rightarrow x$ **then** $f(x_1), f(x_2), f(x_3), \ldots \rightarrow f(x)$

Deduce that $f(x) = a^x$ for *each* number x.

3   Let $a > 0$. Show that for each pair of positive numbers x and y we have

$$\log_a(xy) = \log_a x + \log_a y \quad \text{and} \quad \log_a\left(\frac{x}{y}\right) = \log_a x - \log_a y$$

4   Imagine that you had two copies of a ruler marked out in a 'logarithmic scale' as shown:

1	2	3	4	5	6	7 8 9 10	15	20

1	2	3	4	5	6	7 8 9 10	15	20

(In this case the number x is marked at a distance proportional to $\log_{10} x$ from the left-hand end.) Explain how you would use these rulers to multiply and divide two numbers. (This was the basis of the 'slide rule'.)

(Solutions on page 249)

Limits of functions

Consider the function

$$f(x) = [x] \quad x \in \mathbb{R}$$

(the familiar 'integer part'). What number (if any) does $f(x)$ approach as x approaches $2\frac{1}{2}$? (I don't care about the value of f at $2\frac{1}{2}$, only of $f(x)$ as x gets closer and closer to $2\frac{1}{2}$.) If you let x get closer and closer to $2\frac{1}{2}$ then as soon as x is in region A in the left-hand figure below the value of $f(x)$ will actually be 2. So in this case it seems clear that '$f(x)$ approaches the limit of 2 as x approaches $2\frac{1}{2}$' and we shall write $\lim_{x \to 2\frac{1}{2}} f(x) = 2$ or $\lim_{x \to 2\frac{1}{2}} [x] = 2$.

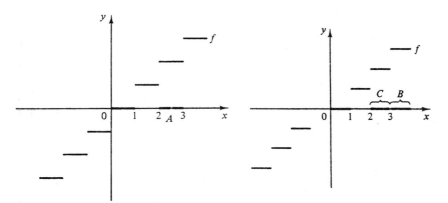

But what happens as x approaches 3? It depends on *how* that approach is made. For if x approaches 3 'from above', i.e. through region B in the right-hand figure above, then $f(x)$ is 3: but if x approaches 3 'from below', i.e. through region C, then $f(x)$ is 2. Indeed in such cases we can choose a rather wobbly approach to 3 such as

$$2\tfrac{1}{2},\ 3\tfrac{1}{3},\ 2\tfrac{3}{4},\ 3\tfrac{1}{5},\ 2\tfrac{5}{6},\ 3\tfrac{1}{7},\ \ldots \to 3$$

but the values that f takes are then

$$2, 3, 2, 3, 2, 3, \ldots$$

which tend to nothing in particular. We shall say that $\lim_{x \to 3} f(x)$ (or $\lim_{x \to 3} [x]$) does not exist.

In general if we wish to consider the behaviour of a function f as x approaches some particular number x_0 then the value of the function *at* x_0 is not considered at all: in fact x_0 need not even be in f's domain. For example if f has domain $(0, \infty)$ then we can consider the behaviour of f as x approaches 0 (because even though 0 is not in f's domain x can certainly approach 0 through f's domain, by taking values $1, \tfrac{1}{2}, \tfrac{1}{3}, \tfrac{1}{4}, \ldots$ for example). However, we certainly would not consider the behaviour of f as x approaches -2 because there is no way for x to approach -2 whilst in f's domain.

Example   Let the function g be given by

$$g(x) = x^2 \quad x \in \mathbb{R}, x \notin \mathbb{Z}$$

i.e. g's domain is all the real numbers which are *not* integers. Does $g(x)$ approach any particular number as x approaches 2?

Solution   The graph of g is as shown

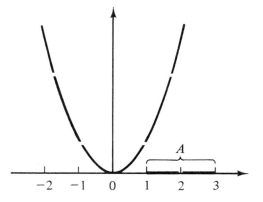

As x approaches 2 it will eventually be in region A and $g(x)$ equals x^2 throughout this region. Hence as x approaches 2 it seems that $g(x)$ must approach 4. We shall call that the 'limit of g as x approaches 2' and write '$\lim_{x \to 2} g(x) = 4$'.   □

We saw the idea of limits of functions in the previous section when we tried to show in an example that 10^x approaches 1 as x approaches 0. In the light of our comments about the value *at 0* being irrelevant the two parts of that example (on page 81) were essentially as follows:

Example   Let $\varepsilon > 0$. Find a $\delta > 0$ such that if $-\delta < x < \delta$ (and $x \neq 0$) then $1 - \varepsilon < 10^x < 1 + \varepsilon$;

i.e. we can ensure that 10^x gets arbitrarily close to 1 by making x sufficiently close to 0.

Example   Show that if $x_1, x_2, x_3, \ldots$ is a sequence which converges to 0 (but none of whose terms equals 0) then the sequence $10^{x_1}, 10^{x_2}, 10^{x_3}, \ldots$ converges to 1; i.e. if $x_1, x_2, x_3, \ldots$ ($\neq 0$) tends to 0 then $10^{x_1}, 10^{x_2}, 10^{x_3}, \ldots$ tends to 1.

To define formally the idea of a function approaching some 'limit' l as x approaches x_0 we could use either of the above ideas, but we shall choose to use sequences as on the right-hand side.

Definition   To say that the function f *converges* to (or *tends* to) a *limit* l as x *converges* to (or *tends* to) x_0 means that

if $x_1, x_2, x_3, x_4, \ldots \to x_0$

(in f's domain and $\neq x_0$)

then $f(x_1), f(x_2), f(x_3), f(x_4), \ldots \to l$

We then write $\lim_{x \to x_0} f(x) = l$.

   (As we commented earlier we shall only ever consider this idea when x_0 is such that at least one sequence (x_n) exists as described in that definition.)

Example   Let the function f be defined by

$$f(x) = \frac{\exp(x) - 1}{x} \qquad x \neq 0$$

where $\exp(x) = 1 + x + x^2/2! + x^3/3! + \cdots$ (as on page 72). Find $\lim_{x \to 0} f(x)$ (if it exists).

Solution   Note firstly that although $f(0)$ is not defined it is certainly possible for x to approach 0 in f's domain. Note also that we showed on page 71 that the infinite sum

$$1 + x + \frac{x^2}{2!} + \frac{x^3}{3!} + \cdots$$

made sense for any number x. Hence

$$f(x) = \frac{\exp(x) - 1}{x} = \frac{x + x^2/2! + x^3/3! + \cdots}{x} = 1 + \frac{x}{2!} + \frac{x^2}{3!} + \frac{x^3}{4!} + \cdots$$

You might then see that $f(x)$ converges to the limit 1 as x converges to 0. But we must formally apply our new definition. Before proceeding note that for each x with $-2 < x < 2$

$$1 - \frac{\frac{1}{2}|x|}{1 - \frac{1}{2}|x|} = 1 - \frac{|x|}{2} - \frac{|x|^2}{2^2} - \cdots \leqslant f(x) = 1 + \frac{x}{2!} + \frac{x^2}{3!} + \cdots$$

$$\leqslant 1 + \frac{|x|}{2} + \frac{|x|^2}{2^2} + \cdots = 1 + \frac{\frac{1}{2}|x|}{1 - \frac{1}{2}|x|}$$

Now to apply the formal definition let

$$\underbrace{x_1, x_2, x_3, \ldots \to 0}_{\neq 0}$$

Then eventually the x_n's will be between -2 and 2 and we can then deduce that

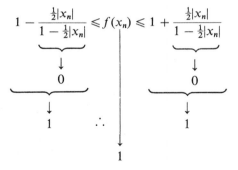

and so

$$f(x_1), f(x_2), f(x_3), \ldots \to 1$$

Hence

if $x_1, x_2, x_3, \ldots \to 0$ then $f(x_1), f(x_2), f(x_3), \ldots \to 1$

$$\underbrace{}_{\neq 0}$$

and so by the definition we have $\lim_{x \to 0} f(x) = 1$. In fact the graph of f is as shown:

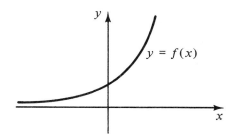

If you were now asked to find

$$\lim_{x \to 0} \frac{\exp(x) + x^2 + x - 1}{x}$$

you wouldn't actually have to do much work because

$$\frac{\exp(x) + x^2 + x - 1}{x} = x + 1 + \frac{\exp(x) - 1}{x}$$

and by the previous example as x tends to 0 this expression will surely tend to $0 + 1 + 1 \ (= 2!)$. In fact it is easy to see that you can add, subtract, multiply and divide limits of functions in a natural way. For example let us consider functions f and g with some suitably-overlapping domains. Suppose that $\lim_{x \to x_0} f(x) = l$ and $\lim_{x \to x_0} g(x) = m$. It follows that:

if $\underbrace{x_1, x_2, x_3, \ldots}_{\text{in } f\text{'s and } g\text{'s domain and } \neq x_0} \to x_0$

(and we're assuming as always that at least one such sequence exists)

then $f(x_1), f(x_2), f(x_3), \ldots \to l$ and $g(x_1), g(x_2), g(x_3), \ldots \to m$

and so by the fact that we can add convergent sequences we can then deduce that

$$f(x_1) + g(x_1), \quad f(x_2) + g(x_2), \quad f(x_3) + g(x_3), \ldots \to l + m$$

Hence, as expected, $\lim_{x \to x_0} (f(x) + g(x)) = l + m$. The subtraction, multiplication and division of limits of functions can similarly be deduced from the corresponding results about sequences.

In the next example we see how to verify that certain limits *don't* exist.

Example   Let the function sgn be defined by

$$\text{sgn}(x) = \begin{cases} 1 & \text{if } x > 0 \\ 0 & \text{if } x = 0 \\ -1 & \text{if } x < 0 \end{cases}$$

Show that $\lim_{x \to 0} \text{sgn}(x)$ does not exist.

Solution   The graph of sgn is shown below.

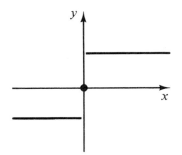

The value of the function at 0 is not relevant to $\lim_{x \to 0} \text{sgn}(x)$. Clearly if x approaches 0 through positive numbers we get quite different values of the function from when x approaches 0 through negative numbers. So a neat way of seeing that $\lim_{x \to 0} \text{sgn}(x)$ doesn't exist is to consider a sequence of positive and negative numbers which approaches 0. For example

$$1, -\tfrac{1}{2}, \tfrac{1}{3}, -\tfrac{1}{4}, \tfrac{1}{5}, \ldots \to 0$$

but

$$\left.\begin{array}{ccccc}
\text{sgn}(1), & \text{sgn}(-\tfrac{1}{2}), & \text{sgn}(\tfrac{1}{3}), & \text{sgn}(-\tfrac{1}{4}), & \text{sgn}(\tfrac{1}{5}), \ldots \\
\| & \| & \| & \| & \| \\
1, & -1, & 1, & -1, & 1 \quad \ldots
\end{array}\right\} \text{doesn't converge}$$

Hence $\lim_{x \to 0} \text{sgn}(x)$ does not exist.   ☐

Before proceeding to the next concept we note without actually giving all the definitions that the idea of the limit

$$\lim_{x \to x_0} f(x) = l$$

extends naturally to

$$\lim_{x \to \infty} f(x) = l \qquad \lim_{x \to x_0} f(x) = \infty \qquad \lim_{x \to \infty} f(x) = -\infty \quad \text{etc.}$$

One example will illustrate the general procedure:

Example   Show that $\lim_{x \to 0} (1/x^2) = \infty$.

Solution   Let f be the function in question; i.e.

$$f(x) = \frac{1}{x^2} \quad x \neq 0$$

Following the general principle of limits we shall show that

$$\text{if } \underbrace{x_1, x_2, x_3, \ldots \to 0}_{\neq 0}$$

$$\textbf{then } f(x_1), f(x_2), f(x_3), \ldots \to \infty$$

So let

$$\underbrace{x_1, x_2, x_3, \ldots \to 0}_{\neq 0}$$

Then

$$\underbrace{x_1{}^2, x_2{}^2, x_3{}^2, \ldots \to 0}_{> 0}$$

and so (by exercise 7 on page 51) the sequence

$$\frac{1}{x_1{}^2}, \frac{1}{x_2{}^2}, \frac{1}{x_3{}^2}, \ldots \to \infty$$

i.e.

$$f(x_1), f(x_2), f(x_3), \ldots \to \infty$$

as required. We have therefore shown that

$$\text{if } \underbrace{x_1, x_2, x_3, \ldots \to 0}_{\neq 0} \quad \textbf{then} \quad f(x_1), f(x_2), f(x_3), \ldots \to \infty$$

which is precisely what we shall mean by $\lim_{x \to 0} f(x) = \infty$. □

Exercises

1   In each of the following cases decide either that the limit is a real number (in which case find it) or ∞ or $-\infty$ or that it does not exist. Justify your answer in each case.

(i) $\lim_{x \to 1} (x - [x])$ (ii) $\lim_{x \to 2} |x - [x] - \frac{1}{2}|$ (iii) $\lim_{x \to 0} \log_{10} x$

2   Let f be the function given by

$$f(x) = \begin{cases} \dfrac{|x|^2 + x|x| - 2}{x - 1} & x \neq 1 \\[2mm] 3 & x = 1 \end{cases}$$

Show that for each $x_0 \in \mathbb{R}$ the limit $\lim_{x \to x_0} f(x)$ exists and that for $x_0 \neq 1$ the limit equals $f(x_0)$ itself.

Sketch the graph of f.

To what number would you change the '3' in the definition of f in order to get a function whose graph does not have a 'break'?

(Solutions on page 252)

Continuous improvements

Consider the following two similar examples:

Example Let

$$f(x) = \begin{cases} \dfrac{x^2 - 1}{x^3 - 1} & x \neq 1 \\[2mm] 0 & x = 1 \end{cases}$$

Find $\lim_{x \to 1} f(x)$ (if it exists).

Example Let

$$g(x) = \begin{cases} \dfrac{x^2 - 1}{x^3 - 1} & x \neq 1 \\[2mm] \frac{2}{3} & x = 1 \end{cases}$$

Find $\lim_{x \to 1} g(x)$ (if it exists).

Solution Since the value of the function at 1 is irrelevant and since these two functions have the same value at each x not equal to 1 it is clear that the two answers are the same. Hence we shall just provide the solution for f.

Now for $x \neq 1$ we have

$$f(x) = \frac{x^2 - 1}{x^3 - 1} = \frac{x + 1}{x^2 + x + 1}$$

and so

if $\underbrace{x_1, x_2, x_3, \ldots}_{\neq 1} \to 1$

then $\left. \begin{array}{cccc} f(x_1), & f(x_2), & f(x_3), & \cdots \\ \| & \| & \| & \\ \dfrac{x_1 + 1}{x_1^2 + x_1 + 1} & \dfrac{x_2 + 1}{x_2^2 + x_2 + 1} & \dfrac{x_3 + 1}{x_3^2 + x_3 + 1} & \cdots \end{array} \right\} \to \frac{2}{3}$

Hence

$$\lim_{x \to 1} f(x) = \lim_{x \to 1} g(x) = \frac{2}{3} \qquad \qquad \square$$

The graphs of the functions f and g from that example are shown below:

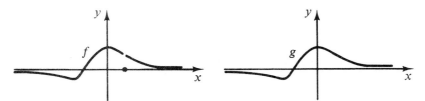

So why does f have a 'break' at $x = 1$ whereas g does not?

For the function f	For the function g
$\lim_{x \to 1} f(x) = \frac{2}{3}$ and $f(1) = 0$	$\lim_{x \to 1} g(x) = \frac{2}{3}$ and $g(1) = \frac{2}{3}$
and so $\lim_{x \to 1} f(x) \neq f(1)$.	and so $\lim_{x \to 1} g(x) = g(1)$.

> **Definition**   Let f be a function and let x_0 be in f's domain. Then to say that f is *continuous at* x_0 means that
>
> if $\underbrace{x_1, x_2, x_3, \ldots}_{\text{in } f\text{'s domain}} \to x_0$ then $f(x_1), f(x_2), f(x_3), \ldots \to f(x_0)$

Otherwise f is said to be *discontinuous* at x_0. The function f is *continuous* if it is continuous at each point in its domain.

Hence a function f is continuous at x_0 if $\lim_{x \to x_0} f(x) = f(x_0)$, where now the value of the function *at* x_0 has become very relevant. As we have seen, a discontinuity is like a 'break' in the function at some point in its domain, and a continuous function is one which has no such breaks.

In the definition there is no longer any need for the sequence (x_n) to avoid x_0. And note that the definition makes sense for any x_0 in f's domain because there will always be a sequence (x_n) in f's domain which converges to x_0 (for example the sequence $x_0, x_0, x_0, \ldots !$).

Example   Let

$$f(x) = \begin{cases} x^2 + 1 & x > 0 \\ -x^2 - 1 & x \leqslant 0 \end{cases}$$

Show that f is discontinuous at $x = 0$ but continuous everywhere else.

Solution   The graph of f illustrates the continuity of f for $x \neq 0$ and the discontinuity (or 'break') at $x = 0$, but we have to formally prove these properties by using the definition.

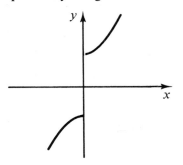

We start by considering the behaviour at $x_0 > 0$. If a sequence converges to x_0 (>0) then eventually (from the Nth term onwards say) all the terms of the sequence will be positive; i.e.

$$x_1, x_2, x_3, \ldots, \underset{>0}{x_N}, \underset{>0}{x_{N+1}}, \underset{>0}{x_{N+2}}, \ldots \to \underset{>0}{x_0}$$

and so, using our established properties about multiplying convergent sequences etc,

$$\left. \begin{array}{ccc} \ldots f(x_N), & f(x_{N+1}), & f(x_{N+2}), \ldots \\ \| & \| & \| \\ x_N^2 + 1, & x_{N+1}^2 + 1, & x_{N+2}^2 + 1, \ldots \end{array} \right\} \to x_0^2 + 1 = f(x_0)$$

We have therefore shown that for $x_0 > 0$

if $x_1, x_2, x_3, \ldots \to x_0$ **then** $f(x_1), f(x_2), f(x_3), \ldots \to f(x_0)$

and so by the definition f is continuous at $x_0 > 0$.

A very similar argument works for $x_0 < 0$.

But what about the behaviour *at* 0? To show that the test for continuity fails we must find a sequence (x_n) which converges to 0 but for which the sequence $(f(x_n))$ does not converge to $f(0)$. There are many ways of doing this but perhaps the easiest is to take a sequence like

$$1, -\tfrac{1}{2}, \tfrac{1}{3}, -\tfrac{1}{4}, \tfrac{1}{5}, -\tfrac{1}{6}, \ldots \to 0$$

and to note that the sequence

$$\underset{>1}{f(1)}, \underset{<-1}{f(-\tfrac{1}{2})}, \underset{>1}{f(\tfrac{1}{3})}, \underset{<-1}{f(-\tfrac{1}{4})}, \underset{>1}{f(\tfrac{1}{5})}, \underset{<-1}{f(-\tfrac{1}{6})}, \ldots$$

doesn't even converge, let alone converge to $f(0)$. Hence we have shown that f is discontinuous at $x = 0$ as required.   □

As that example illustrates again, a discontinuity of a function is often a break in the graph of that function **at some point in the function's domain**. Consider the function f given by

$$f(x) = \frac{1}{x - 1} \quad x \neq 1$$

In the graph of f there is a huge break at $x = 1$ but since 1 is not in the domain of f we cannot talk about the continuity or discontinuity of f there. In fact f is continuous at each point in its domain, and so it is 'continuous'.

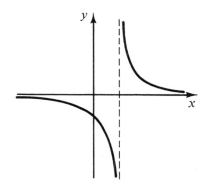

We have actually met the idea of continuity (in disguise) twice before in this book. In chapter 2 we defined some sequences by 'recurrence' and investigated their limits. For example in exercise 2 on page 62 we considered the sequence defined by

$$x_1 = 10 \quad \text{and} \quad x_n = f(x_{n-1}) \quad \text{for} \quad n > 1$$

where

$$f(x) = \frac{x}{2} + \frac{1}{x}$$

We proved that

$$x_1, x_2, x_3, \ldots \to x_0$$

where x_0 is some positive number. We then used properties of sequences to deduce that

$$f(x_1), f(x_2), f(x_3), \ldots \to f(x_0)$$

and then to find x_0. In other words we were using the continuity of f. However when trying the same process (on page 61) with

$$x_1 = 1 \quad \text{and} \quad x_n = g(x_{n-1}) \quad \text{for} \quad n > 1$$

where

$$g(x) = \frac{x}{2} + \frac{1}{[x]}$$

we found that the process went wrong because

$$x_1, x_2, x_3, \ldots \to 2$$

but

$$g(x_1), g(x_2), g(x_3), \ldots \not\to g(2)$$

With hindsight we can now see that this is because g is not continuous at $x = 2$.

In general we can define a sequence (x_n) by a recurrence relation of the form

$$x_1 = a \quad \text{and} \quad x_n = f(x_{n-1}) \quad \text{for} \quad n > 1$$

If the sequence converges to x_0 **and** the function f is continuous at x_0 then the limit x_0 will actually satisfy $x_0 = f(x_0)$ (x_0 is a 'fixed point' of f). We shall prove this in the next exercises and show that it is a very powerful method for obtaining approximate solutions to equations.

The other place where we have met continuity before was when we extended the definition of 10^x from the rational numbers to all the real numbers. On page 81 we drew the graph of

$$f(x) = 10^x \quad x \in \mathbb{Q}$$

and then extended the domain to $\mathbb{R}$ by 'filling in the gaps' in a natural way.

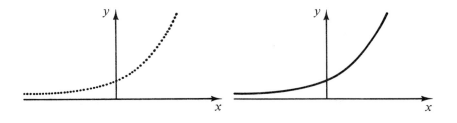

With our new concept of continuity we can now see that we were merely choosing the new values of 10^x so that the function

$$f(x) = 10^x \quad x \in \mathbb{R}$$

is continuous.

In the same way exercise 2 on page 88 is then saying, for example, that any continuous function f with domain $\mathbb{R}$, $f(1) = 10$ and

$$f(x + y) = f(x) \times f(y)$$

for each x and y must be the function $f(x) = 10^x$.

Since the function

$$f(x) = a^x \quad x \in \mathbb{R}$$

is continuous (which we think of informally as having no break at any point of its domain) and since the graph of the function

$$f^{-1}(x) = \log_a x \quad x > 0$$

is just a mirror-image of that of f, surely f^{-1} too will be continuous? In fact surprisingly we only use the fact that f is strictly increasing to deduce that f^{-1} is continuous.

> ***Theorem***   Let f be any strictly increasing function whose domain D is an interval. Then the function f^{-1} is continuous.
> (A similar result holds for the inverse of a strictly decreasing function.)

(Before proceeding to prove the theorem we illustrate two typical functions to which the theorem applies:

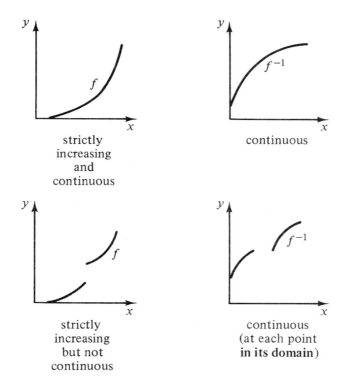

strictly
increasing
and
continuous

continuous

strictly
increasing
but not
continuous

continuous
(at each point
in its domain)

Of course the case of a^x and $\log_a x$ is as in the left-hand pair where f itself is continuous.)

> ***Proof of the theorem***   Let f be strictly increasing with domain D, which is an interval, and range G. Then, as we saw in exercise 2 on page 79, f has an inverse function f^{-1} which is strictly increasing with domain G and range D.

To show that f^{-1} is continuous at an arbitrary point x_0 in its domain G we have to consider any sequence

$$\underbrace{x_1, x_2, x_3, \ldots}_{\in G} \to x_0$$

and show that

$$f^{-1}(x_1), \quad f^{-1}(x_2), \quad f^{-1}(x_3), \ldots \to f^{-1}(x_0)$$

To show that this latter sequence converges we'll go back to the basic definition of convergence of a sequence. Let $\varepsilon > 0$. We have to find an N with

$$f^{-1}(x_0) - \varepsilon < f^{-1}(x_n) < f^{-1}(x_0) + \varepsilon \quad \text{for all } n \geqslant N$$

But $f^{-1}(x_0) \in D$ which is an interval. Let us assume that $f^{-1}(x_0) - \varepsilon$ and $f^{-1}(x_0) + \varepsilon$ are both in D as shown: (if they're not the proof would merely need minor adjustments).

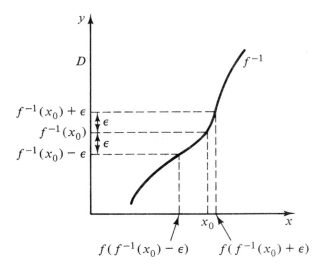

Then

$$f^{-1}(x_0) - \varepsilon < f^{-1}(x_0) < f^{-1}(x_0) + \varepsilon$$

and as f is strictly increasing we can apply f right through these inequalities without upsetting the order and deduce that

$$f(f^{-1}(x_0) - \varepsilon) < \underbrace{f(f^{-1}(x_0))}_{= x_0} < f(f^{-1}(x_0) + \varepsilon)$$

as illustrated.

Now since the sequence (x_n) converges to x_0 it follows that for some N the terms $x_N, x_{N+1}, x_{N+2}, \ldots$ are all between $f(f^{-1}(x_0) - \varepsilon)$ and $f(f^{-1}(x_0) + \varepsilon);$

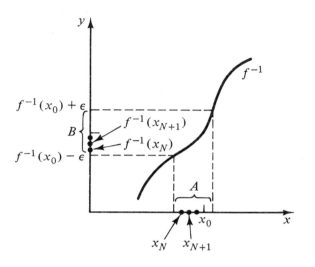

i.e.

$$\left. \begin{array}{l} f(f^{-1}(x_0) - \varepsilon) < x_N < f(f^{-1}(x_0) + \varepsilon) \\ f(f^{-1}(x_0) - \varepsilon) < x_{N+1} < f(f^{-1}(x_0) + \varepsilon) \\ f(f^{-1}(x_0) - \varepsilon) < x_{N+2} < f(f^{-1}(x_0) + \varepsilon) \\ \vdots \end{array} \right\} \text{ as shown in } A$$

But f^{-1} is strictly increasing and so we can apply it right through these inequalities without upsetting the order to deduce that

$$\left. \begin{array}{l} f^{-1}(x_0) - \varepsilon < f^{-1}(x_N) < f^{-1}(x_0) + \varepsilon \\ f^{-1}(x_0) - \varepsilon < f^{-1}(x_{N+1}) < f^{-1}(x_0) + \varepsilon \\ f^{-1}(x_0) - \varepsilon < f^{-1}(x_{N+2}) < f^{-1}(x_0) + \varepsilon \\ \vdots \end{array} \right\} \text{ as shown in } B$$

But that's exactly what we needed to show that the sequence

$$f^{-1}(x_1), \quad f^{-1}(x_2), \quad f^{-1}(x_3), \ldots \to f^{-1}(x_0)$$

Hence

if $x_1, x_2, x_3, \ldots \to x_0$ **then** $f^{-1}(x_1), f^{-1}(x_2), f^{-1}(x_3), \ldots \to f^{-1}(x_0)$

and we have shown that f^{-1} is continuous at any x_0 in its domain; i.e. f^{-1} is continuous. $\square$

So the function $\log_a$, being the inverse of the function a^x, is continuous as expected. There are many other ways of constructing 'new' continuous functions from 'old' ones. For example our earlier comments about adding, subtracting, multiplying and dividing limits means that the sum, difference, product and quotient of continuous functions are themselves continuous. For instance if f and g are both continuous at x_0 then

$$\text{if } \underbrace{x_1, x_2, x_3, \ldots}_{\text{in } f\text{'s and } g\text{'s domain}} \to x_0$$

then

$$f(x_1), f(x_2), f(x_3), \ldots \to f(x_0)$$

and

$$g(x_1), g(x_2), g(x_3), \ldots \to g(x_0)$$

and so by our result about adding sequences

$$\text{then } f(x_1) + g(x_1), f(x_2) + g(x_2), f(x_3) + g(x_3), \ldots \to f(x_0) + g(x_0)$$

Hence the function $f + g$ (i.e. the function whose value at x is $f(x) + g(x)$) is also continuous at x_0. Similarly for the difference, product and quotient of two functions. The domain of the new function will consist of those numbers in the domains of both f and g (i.e. it will be the 'intersection' of their domains) and in the case of the quotient f/g the domain must also exclude the points where $g(x) = 0$.

Another way of combining continuous functions to get other (in general much more complicated) continuous functions is by looking at the composite 'function of a function'. For example if $f(x) = \log_{10} x$ and $g(x) = x - 1$ then the *composite* function $f \circ g$ is the one whose value at x is $f(g(x))$ or $\log_{10}(x - 1)$. But what are the domains of these functions?

$$f(x) = \log_{10} x \quad (x > 0) \qquad g(x) = x - 1 \quad (x \in \mathbb{R})$$
$$f \circ g(x) = \log_{10}(x - 1) \quad (x \in \text{???})$$

Mathematicians differ about the domain of the composite function. Some purists would say that we cannot even define $f \circ g$ in this case because, for example, although $g(0)$ is defined $f \circ g(0)$ is not. Others take the more pragmatic view that the domain of $f \circ g$ is the set of all those x for which $f(g(x))$ makes sense; i.e. those x in g's domain for which $g(x)$ is in f's domain. We shall take this latter more practical approach. In this case therefore we require that $x - 1$ is positive and then the composite function $f \circ g$ has domain $\{x \in \mathbb{R} : x > 1\}$. It turns out that compositions of

continuous functions are themselves continuous, and the next example
illustrates this and the other techniques outlined above.

Example Show that the following functions are continuous.

(i) $g(x) = x^2 + 1$ $(x \in \mathbb{R})$

(ii) $h(x) = 2^{x^2+1}$ $(x \in \mathbb{R})$

(iii) $j(x) = \dfrac{x2^{x^2+1} - \sqrt{x}}{\log_2 x + 1}$ $(x > 0 \text{ and } x \neq \tfrac{1}{2})$

Solution (i) We start by showing the obvious fact that the
functions k and l given by

$$k(x) = c \text{ (a constant)} \quad \text{and} \quad l(x) = x$$

are continuous: for trivially

$$
\begin{array}{c}
\textbf{if} \qquad x_1, \quad x_2, \quad x_3, \quad \ldots \to \quad x_0 \\
\textbf{then(!)} \quad k(x_1), \ k(x_2), \ k(x_3), \ \ldots \to \ k(x_0) \\
\quad \| \qquad \| \qquad \| \qquad \qquad \| \\
\quad c \qquad c \qquad c \qquad \qquad c
\end{array}
$$

and

$$
\begin{array}{c}
l(x_1), \ l(x_2), \ l(x_3), \ \ldots \to \ l(x_0) \\
\| \qquad \| \qquad \| \qquad \qquad \| \\
x_1 \qquad x_2 \qquad x_3 \qquad \qquad x_0
\end{array}
$$

and so k and l are continuous at each x_0 (and hence continuous).

Now g is obtained from those trivial functions by multiplying l by itself
and adding k (with the constant equal to 1). So by the previous comments
about adding and multiplying continuous functions it follows that g is
continuous.

(Of course it is easy to check directly that g is continuous because **if** the
sequence (x_n) converges to x_0 **then** the sequence $(x_n^2 + 1)$ converges to
$x_0^2 + 1$.)

(ii) Let $f(x) = 2^x$ (which we know is continuous) and $g(x) = x^2 + 1$ (as
in the previous part). Then

$$h(x) = f(x^2 + 1) = f(g(x))$$

and h is the composite function $f \circ g$. Now

$$\textbf{if} \qquad x_1, \quad x_2, \quad x_3, \quad \ldots \to \quad x_0$$

then (by the continuity of g at x_0)

$$
\begin{array}{c}
x_1^2 + 1, \ x_2^2 + 1, \ x_3^2 + 1, \quad \ldots \to \quad x_0^2 + 1 \\
\| \qquad\quad \| \qquad\quad \| \qquad\qquad\qquad \| \\
g(x_1) \qquad g(x_2) \qquad g(x_3) \qquad\qquad\qquad g(x_0)
\end{array}
$$

and (by the continuity of f at $g(x_0)$)

$$\textbf{then} \begin{cases} 2^{x_1{}^2+1}, & 2^{x_2{}^2+1}, & 2^{x_3{}^2+1}, & \cdots \rightarrow & 2^{x_0{}^2+1} \\ \| & \| & \| & & \| \\ f(g(x_1)) & f(g(x_2)) & f(g(x_3)) & & f(g(x_0)) \\ \| & \| & \| & & \| \\ h(x_1) & h(x_2) & h(x_3) & & h(x_0) \end{cases}$$

Therefore h is continuous at each x_0 and hence it is continuous.

(A similar verification will work for any composition of continuous functions as we shall see in the next exercises.)

(iii) As above the function r given by $r(x) = x^2 \ (x \geqslant 0)$ is continuous: in fact its domain is an interval and it's strictly increasing. Hence by the theorem on page 101 its inverse function $s(x) = \sqrt{x}$ is also continuous. Therefore

$$j(x) = \frac{x2^{x^2+1} - \sqrt{x}}{\log_2 x + 1}$$

is a simple arithmetic $(+, -, \times, \div)$ combination of continuous functions and it is therefore continuous. (Its domain is the common domain of those contributing functions where we must also exclude points where the bottom of the fraction is zero.)   □

We now have the necessary general concepts of limits and continuity and, after some exercises, we can return to our study of some specific functions.

Exercises

·1   Suppose that f and g are functions with g continuous at some point x_0 and f continuous at $g(x_0)$. Show that the composite function $f \circ g$ is continuous at x_0.

Show that the function $h(x) = [\log_2 x]$ – the integer part of the $\log_2 x$ – is continuous at $x = 3$.

2   Let n be a positive integer and let r be any rational number. Show that the following functions are continuous:

(i) $f(x) = x^n \quad (x \in \mathbb{R})$
(ii) $g(x) = x^{-n} \quad (x \neq 0)$
(iii) $h(x) = \sqrt[n]{x} \quad (x \geqslant 0)$
(iv) $j(x) = x^r \quad (x > 0)$

3   Let f be a continuous function with domain D and range contained in D, and let $a \in D$. Suppose that we define a sequence (x_n) by the recurrence relation

$$x_1 = a \quad \text{and} \quad x_n = f(x_{n-1}) \quad \text{for} \quad n > 1$$

and that the sequence converges to x_0. Show that $x_0 = f(x_0)$.
   Note that the function f given by

$$f(x) = \frac{x^3 + 2x^2 + 1}{8} \quad x \in \mathbb{R}$$

is continuous. Let $x_1 = 0$ and use f to define a sequence (x_n) in the above way. Use your calculator to work out the first few terms of this sequence and hence find, to three decimal places, a root of the equation

$$x^3 + 2x^2 - 8x + 1 = 0$$

4   Assume that f is a strictly increasing function with domain $(0, \infty)$ and range $\mathbb{R}$ which satisfies

(i) $f(a) = 1$
(ii) $f(x \times y) = f(x) + f(y)$ for each $x, y > 0$

Show that f^{-1} is a continuous function which satisfies

(i)′ $f^{-1}(1) = a$
(ii)′ $f^{-1}(x + y) = f^{-1}(x) \times f^{-1}(y)$ for each $x, y \in \mathbb{R}$

Deduce that $f(x) = \log_a x$ for each $x > 0$.

5   Let f be a continuous function whose domain includes the interval $[a, b]$ and let $x_0 \in (a, b)$ have $f(x_0) > 0$. Show that f is positive throughout some interval surrounding x_0.

(Solutions on page 253)

Exp and log

We earlier defined a function exp by

$$\exp(x) = 1 + x + \frac{x^2}{2!} + \frac{x^3}{3!} + \cdots$$

which we shall call the *exponential* function. That name implies that it is something to do with powers and indeed at school you probably called this function 'e^x', where

$$e = \exp(1) = 1 + \frac{1}{1!} + \frac{1}{2!} + \frac{1}{3!} + \cdots$$

is the number we met in chapter 1. But before we can use that name we must show that exp has the following remarkable property: for each number x

$$\underbrace{1 + x + \frac{x^2}{2!} + \frac{x^3}{3!} + \cdots}_{\exp(x)} = \underbrace{\left(1 + \frac{1}{1!} + \frac{1}{2!} + \frac{1}{3!} + \cdots \right)^x}_{e^x}$$

Theorem $\exp(x) = e^x$ for each $x \in \mathbb{R}$.

Proof   It follows from exercise 2 on page 88 that if a function f with domain $\mathbb{R}$ satisfies:

(i) $f(1) = e$;

(ii) $f(x) \times f(y) = f(x + y)$ for each x and y;

(iii) f is continuous;

then $f(x) = e^x$ for each x. So we shall establish that the function exp has these three properties, the first being immediate since we have already noted that $\exp(1) = e$.

We shall show next that exp has the multiplicative property (ii), namely that

$$\exp(x) \times \exp(y) = \exp(x + y)   \text{for each } x \text{ and } y$$

So let x and y be any two numbers. Then

$$\exp(x) = 1 + x + \frac{x^2}{2!} + \frac{x^3}{3!} + \cdots   \text{and}$$

$$\exp(y) = 1 + y + \frac{y^2}{2!} + \frac{y^3}{3!} + \cdots$$

We have never multiplied two infinite sums (or series) together and it takes some justification. Strictly we should think of each of these numbers as a limit of a sequence of partial sums and work out the products of those partial sums:

$$1 \times 1$$

$$(1 + x) \times (1 + y)$$

$$\left(1 + x + \frac{x^2}{2!} \right) \times \left(1 + y + \frac{y^2}{2!} \right)$$

$$\left(1 + x + \frac{x^2}{2!} + \frac{x^3}{3!} \right) \times \left(1 + y + \frac{y^2}{2!} + \frac{y^3}{3!} \right)$$

$$\vdots$$

$$\downarrow$$

$$\exp(x) \times \exp(y)$$

And does this give the same result as multiplying out the brackets

$$\left(1 + x + \frac{x^2}{2!} + \frac{x^3}{3!} + \cdots\right) \times \left(1 + y + \frac{y^2}{2!} + \frac{y^3}{3!} + \cdots\right)$$

in a sensible sort of way? Luckily yes. For example if you multiplied out this last pair of brackets in a common-sense sort of way how many x^2y^3s would you get? $1/(2! \times 3!)$ and that's exactly the number you would have got in any of those products of the partial sums from the fourth product onwards.

So we calculate the product in a straightforward way and obtain the answer shown on the left:

$$\exp(x) \times \exp(y) =$$
$$1 +$$
$$(x + y) +$$
$$\frac{1}{2!}(x^2 + 2xy + y^2) +$$
$$\frac{1}{3!}(x^3 + 3x^2y + 3xy^2 + y^3) +$$
$$\vdots$$

$$\exp(x + y) =$$
$$1 +$$
$$(x + y) +$$
$$\frac{1}{2!}(x + y)^2 +$$
$$\frac{1}{3!}(x + y)^3 +$$
$$\vdots$$

As you can see the answer agrees with that on the right which is the series for $\exp(x + y)$. Hence

$$\exp(x) \times \exp(y) = \exp(x + y)$$

as required.

We still need to show that the function exp is continuous and we do this in two stages, showing firstly that it is continuous at 0. Rather as we have seen before, for $-2 < x < 2$

$$\frac{1 - \frac{3}{2}|x|}{1 - \frac{1}{2}|x|} = 1 - |x| - \frac{|x|^2}{2} - \frac{|x|^3}{2^2} - \cdots$$

$$\leqslant 1 + x + \frac{x^2}{2!} + \frac{x^3}{3!} + \cdots = \exp(x)$$

$$\leqslant 1 + |x| + \frac{|x|^2}{2} + \frac{|x|^3}{2^2} + \cdots = \frac{1 + \frac{1}{2}|x|}{1 - \frac{1}{2}|x|}$$

and hence for $-2 < x < 2$ we have

$$\frac{1 - \frac{3}{2}|x|}{1 - \frac{1}{2}|x|} \leqslant \exp(x) \leqslant \frac{1 + \frac{1}{2}|x|}{1 - \frac{1}{2}|x|}$$

So if $x_1, x_2, x_3, \ldots \to 0$ we have that the x_ns are eventually between -2 and 2 and we can then deduce

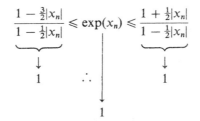

$$\frac{1 - \frac{3}{2}|x_n|}{1 - \frac{1}{2}|x_n|} \leqslant \exp(x_n) \leqslant \frac{1 + \frac{1}{2}|x_n|}{1 - \frac{1}{2}|x_n|}$$

and **then** $\exp(x_1), \exp(x_2), \exp(x_3), \ldots \to 1 = \exp(0)$. This shows that the function exp is continuous at 0. To show that it is continuous at any x_0 note that

if $x_1,$ $x_2,$ $x_3,$ $\ldots \to x_0$

then $x_1 - x_0,$ $x_2 - x_0,$ $x_3 - x_0,$ $\ldots \to 0$

and by the continuity of exp at 0 we deduce that

$$\exp(x_1 - x_0), \quad \exp(x_2 - x_0), \quad \exp(x_3 - x_0), \ldots \to \exp(0) = 1$$

Now by the multiplicative property established above we know that

$$\exp(x_0) \times \exp(x_n - x_0) = \exp(x_0 + (x_n - x_0)) = \exp(x_n)$$

Therefore multiplying through the above sequence by $\exp(x_0)$ we have

then $\exp(x_1), \quad \exp(x_2), \quad \exp(x_3), \ldots \to \exp(x_0)$

and this shows that exp is continuous at any x_0.

So the function exp does satisfy all the properties:

(i) $\exp(1) = e$;
(ii) $\exp(x) \times \exp(y) = \exp(x + y)$ for each x and y;
(iii) exp is continuous;

and this shows that $\exp(x) = e^x$ for each number x.   □

That is just one of the remarkable properties of the exponential function. Another concerns differentiation and will have to wait for a full discussion until the next chapter. But the following example previews that result:

Example   Show that the gradient of the graph of the exponential function at $x = 0$ (i.e. at the point $(0, 1)$) is 1. Deduce that the gradient at (x_0, e^{x_0}) is e^{x_0}.

Solution   Consider the chord on the graph of the exponential function from the point P $(0, 1)$ to any other point Q (x, e^x). Then

$$\text{gradient of } PQ = \frac{e^x - 1}{x - 0}$$

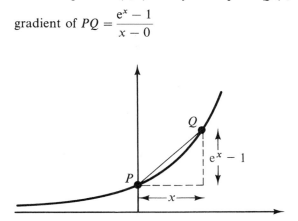

So the gradient of the graph at P (which we touched upon before in our discussion of sequences) can be found by letting Q get closer and closer to P; i.e. by letting x tend to 0 in the above expression. But as we saw in the example on page 91

$$\lim_{x \to 0} \frac{e^x - 1}{x} = 1$$

and so it seems (informally at least) that the exponential function has gradient 1 at the point $(0, 1)$.

In general if P is the point (x_0, e^{x_0}) and Q is any other point (x, e^x) then

$$\text{gradient of } PQ = \frac{e^x - e^{x_0}}{x - x_0} = e^{x_0}\left(\frac{e^{x - x_0} - 1}{(x - x_0) - 0}\right)$$

So as Q approaches P we let x approach x_0 and consider what happens to that expression. But then $x - x_0$ approaches 0 and so by the first part of the example

$$\text{gradient of } PQ = e^{x_0}\left(\frac{e^{x - x_0} - 1}{(x - x_0) - 0}\right) \to e^{x_0} \times 1 = e^{x_0}$$

So it seems (informally at least) that the gradient of the exponential function at (x_0, e^{x_0}) is e^{x_0}. □

We shall confirm that property more formally when we deal with differentiation in the next chapter. It says that if you take any point (x, y) on the graph of the exponential function then the gradient there is just y. The same property will be true for the function $f(x) = e^{x+1}$ or indeed for the function $g(x) = e^{x + 3\frac{1}{4}}$. But it turns out (as we shall see later) that the exponential is the only function with that property and which equals 1 at $x = 0$.

We can summarise these delightful properties of the exponential function as follows:

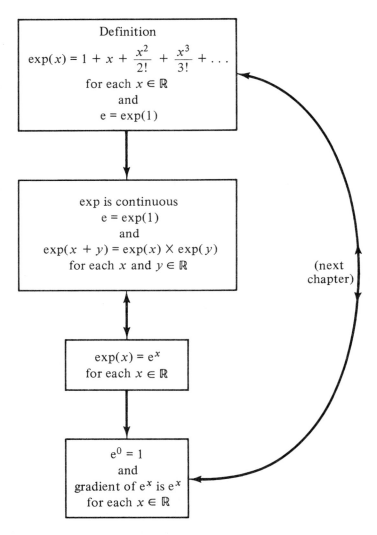

The arrows show the route that we have taken in introducing and developing the exponential function. In fact it is possible as a completely alternative approach to *define* the exponential function as the unique function which satisfies the properties in the last box and to derive the other properties in a different order.

We now turn our attention to the function log (or ln) defined on pages 46 and 47 as follows:

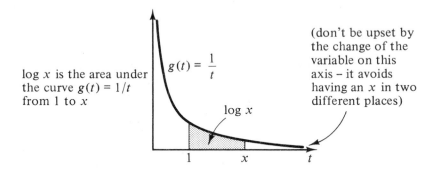

log x is the area under the curve $g(t) = 1/t$ from 1 to x

(don't be upset by the change of the variable on this axis – it avoids having an x in two different places)

That definition is for any positive number x with the understanding that if $x < 1$ then, as you have to travel backwards to measure the area, that area is regarded as negative; e.g.

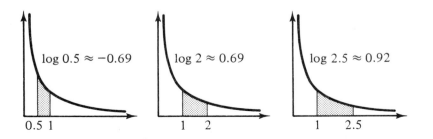

Two properties of log are immediate from the definition: it has domain $(0, \infty)$ and it is strictly increasing, as the typical examples above illustrate. It also seems plausible that areas of all values are possible; i.e. that the function log has range $\mathbb{R}$.

You will know from your school studies that there is a strong link between the log and exponential functions (a link which we have almost presumed by our use of the name 'log') but again we shall have to deduce this link from our definition of the function. Since you will have seen the following result before you may not now appreciate its magnificence: we have defined two natural functions in completely independent ways and yet they turn out to be the closest of soul-mates. Actually we won't have quite enough information to prove the result until we have introduced calculus, but it's unreasonable to have to wait so long for such a gem:

> ***Theorem***   The function log is the inverse of the exponential function.

> ***Sketch proof (or stretch proof?)***   We know that in general the inverse function of a^x is $\log_a x$. Hence the inverse function of e^x is $\log_e x$ and so we have to show that $\log x$ actually coincides with $\log_e x$.

It follows from exercise 4 above that if a strictly increasing function f has domain $(0, \infty)$ and range $\mathbb{R}$ and satisfies

(i) $f(xy) = f(x) + f(y)$ for each $x, y > 0$
(ii) $f(e) = 1$

then f is the function $\log_e$. So to show that our function log is simply $\log_e$ we shall give a sketchy (stretchy) 'proof' that log satisfies (i) and (ii).

To prove (i) we are going to make a surprising diversion into matrices and transformations of the plane (although you don't have to bring matrices into it).

Imagine that I take each point (a, b) in the plane and move it to the point $(3a, b)$. The effect of this 'transformation' on a given shape is shown below:

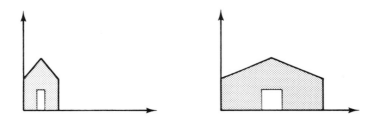

It stretches the shape three-fold in the horizontal direction and so multiplies its area by 3. In fact as you will probably have seen before you can represent this transformation by the two-by-two matrix

$$\begin{pmatrix} 3 & 0 \\ 0 & 1 \end{pmatrix} \quad \text{since} \quad \begin{pmatrix} 3 & 0 \\ 0 & 1 \end{pmatrix}\begin{pmatrix} a \\ b \end{pmatrix} = \begin{pmatrix} 3a \\ b \end{pmatrix}$$

(The fact that areas are multiplied by 3 is connected with the fact that the matrix has 'determinant' 3, but again we don't have to bring determinants into it.)

Consider now the more adventurous transformation which takes each point (a, b) to the point $(3a, \frac{1}{3}b)$: this has matrix

$$\begin{pmatrix} 3 & 0 \\ 0 & \frac{1}{3} \end{pmatrix}$$

and its effect on the same shape is shown below

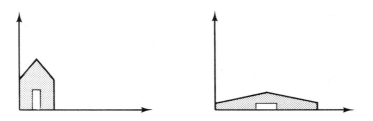

So this transformation stretches shapes three-fold in the horizontal direction and divides them by 3 vertically, thus leaving areas unaltered. More generally let x be any positive number and consider the effect of the transformation which takes each point (a, b) to the point $(xa, b/x)$. This multiplies by x in the horizontal direction and divides by x in the vertical direction, again leaving the area of any shape unchanged. (In fact the matrix of this transformation is

$$\begin{pmatrix} x & 0 \\ 0 & 1/x \end{pmatrix}$$

and it leaves areas unchanged as it has determinant 1.)

But what on earth has this got to do with the new function log? Consider the effect of that last transformation on the area shown on the left below:

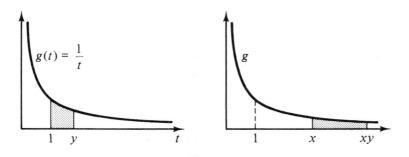

A little thought will show you that it transforms to the area on the right.

But the area on the left is log y and the area on the right is the one which represents log xy but with the area for log x rubbed out. Hence by a slightly sketchy diversion we have been able to see that

$$\log y = \log xy - \log x \quad \text{and} \quad \log xy = \log x + \log y$$

So in trying to establish the two properties

(i) $f(xy) = f(x) + f(y)$ for each $x, y > 0$
(ii) $f(e) = 1$

for log all that remains to prove is that log e $= 1$. But in exercise 2 on page 28 we constructed a sequence converging to log e and saw that the limit seemed to be 1.

Our 'proof' has been a little sketchy but at least it has enabled us to see the stunning link between the two functions exp and log.   □

It follows from that result that the function log is just the logarithmic function $\log_e$, thus justifying its name: we can now feel free to call it the *natural logarithm* function, sometimes denoted by ln. Although we were able to state the theorem in simple terms it is saying something quite complicated. For example $\log(\exp(x)) = x$ means the following:

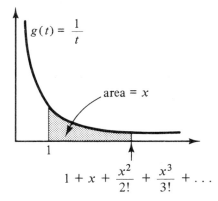

Earlier we summarised the properties of the exponential function and now we can extend that summary to include the log function. Again the arrows show our route in deriving the properties (some of which will have to wait until the next chapter on differentiation). The dotted line shows that our proof was only sketchy but as you'll see there will eventually be a foolproof route around this set of properties.

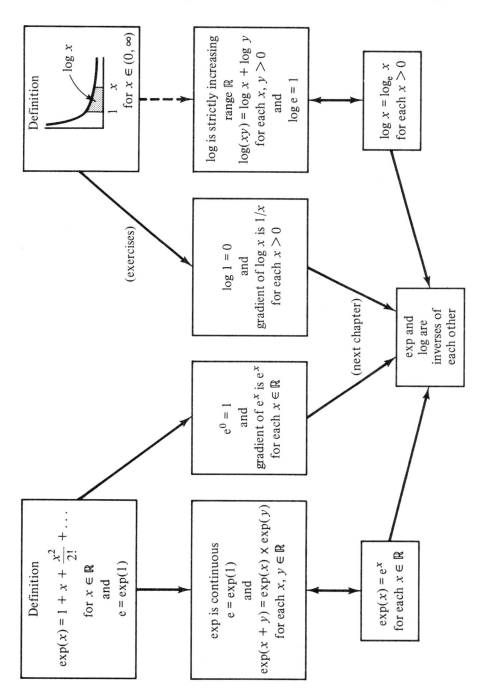

These properties are so inter-linked that it is possible to introduce the functions by practically any pair of the properties and then to deduce the rest. The marriage of these two functions is one of the great thrills of elementary analysis. Surely no budding analyst can fail to have been moved/surprised/delighted/intrigued by their development?

Exercises

1   Show that the function $f(x) = e^{-x}$ ($x \in \mathbb{R}$) is continuous.

Let $x_1 = 0$ and $x_n = f(x_{n-1})$ for $n > 1$ (as in exercise 3 on page 107). Calculate the first few terms of the sequence (x_n) and hence find, to three decimal places, a root of the equation $x + \log x = 0$.

Sketch the graphs $y = \log x$ and $y = -x$ on the same axes and observe that the equation $x + \log x = 0$ has a unique solution.

2   Use the fact that $e^{\alpha\beta} = (e^\alpha)^\beta$ to show that

(i) $a^x = e^{x \log a}$ for each positive number a and each number x;

(ii) $\log_a x = \log x / \log a$ for each positive a and x with $a \neq 1$.

3   (i) Let x and x_0 be two different positive numbers. By referring to the shaded area illustrated below show that

$$(x - x_0)\frac{1}{x} < \log x - \log x_0 < (x - x_0)\frac{1}{x_0}$$

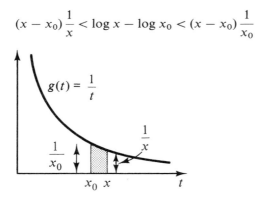

Deduce that

$$\lim_{x \to x_0} \frac{\log x - \log x_0}{x - x_0} = \frac{1}{x_0}$$

(ii) Let P be the point $(x_0, \log x_0)$ on the graph of the log function and let Q be any other point $(x, \log x)$ on that graph. Find in terms of x_0 and x the gradient of the chord PQ. Show that this gradient tends to $1/x_0$ as Q approaches P. (We are therefore tempted to say that the gradient of the log function at P equals $1/x_0$.)

(Solutions on page 256)

Going round in circles

Three of the functions which you met early at school are the trigonometric functions, sine, cosine and tangent. You will have first encountered
them as useful ratios in a right-angled triangle and at that stage the
angles will have been measured in degrees. So we shall start by defining
these functions in terms of degrees and we shall also start with the common convention of denoting an angle by the greek letter θ ('theta').

Imagine that you start at the point $(1, 0)$ in the coordinate plane and
move anti-clockwise in a circle around the origin, going through an angle
of $\theta°$ ending up at the point C. Then θ can take any value, and three typical
examples are illustrated:

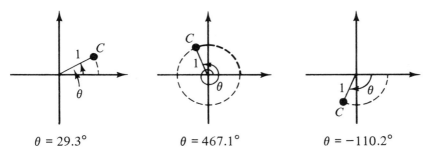

$\theta = 29.3°$ $\theta = 467.1°$ $\theta = -110.2°$

Now the trigonometric functions are defined by

$\cos \theta$ = the x-coordinate of C after that turn of θ

$\sin \theta$ = the y-coordinate of C after that turn of θ

$\tan \theta = \sin \theta/\cos \theta$ (provided that $\cos \theta \neq 0$)

So in the case of an acute angle of $\theta°$ ($0 \leqslant \theta < 90$) in a right-angled triangle
these functions are precisely as expected from school trigonometry:

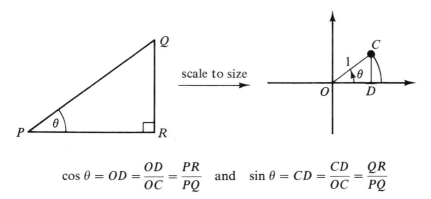

scale to size

$$\cos \theta = OD = \frac{OD}{OC} = \frac{PR}{PQ} \quad \text{and} \quad \sin \theta = CD = \frac{CD}{OC} = \frac{QR}{PQ}$$

Our definition of the trigonometric functions enables us to prove all the usual identities linking the functions. For example Pythagoras' theorem in the triangle *OCD* below gives

$$CD^2 + OD^2 = OC^2$$

or

$$(\sin \theta)^2 + (\cos \theta)^2 = 1$$

which we usually write as

$$\sin^2 \theta + \cos^2 \theta = 1$$

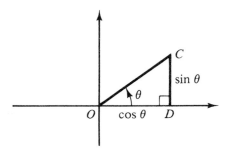

Similarly, completing a turn of $\theta°$ to the point *C* and then a further turn of $90°$ to *C'* gives the situation illustrated:

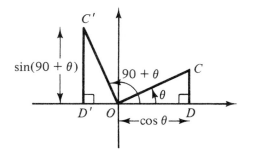

It's easily seen that the angle *OC'D'* is also $\theta°$ and hence that triangle *OC'D'* is congruent to triangle *COD*. Therefore $C'D' = OD$ and $\sin(90 + \theta) = \cos \theta$; etc. We shall now assume that all the usual trigonometric identities hold.

It is clear from the definition of the trigonometric functions that the values repeat themselves every time the point C completes a full circle, or every 360°. These functions are said to have *period* 360° and are called *periodic*. It is also clear that sine and cosine repeatedly take values between -1 and 1 giving rise to the wave-like graphs illustrated. (Note that we shall now switch to x as the variable, more in line with the functions in previous sections. Note also that we have to condense the values along the x-axis in order to get a good view of the periodic behaviour of these functions.)

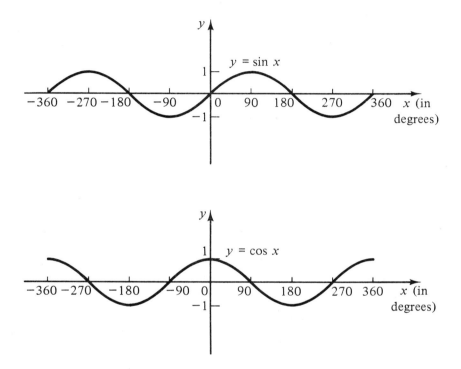

Before proceeding to an analytical study of these functions we must decide whether to continue using degrees or whether to measure angles in some other way better-suited to our analysis. The 360° in a full revolution is a left-over from Babylonian arithmetic in which 'base' 60 featured (leaving other relics like the 60 minutes in an hour). Surely it's time that we decimalised the angle (with 100 'deci-degrees' to a full revolution?) or introduced some other measure of angles (with k units to a full revolution for some suitably-chosen number k)? The effect on the graphs of the trigonometric functions would simply be to change the scale on the x-axis, as shown below for the sine function:

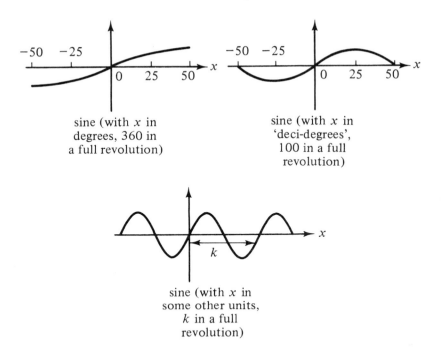

sine (with x in
degrees, 360 in
a full revolution)

sine (with x in
'deci-degrees',
100 in a full
revolution)

sine (with x in
some other units,
k in a full
revolution)

In fact we're going to choose k to be approximately 6.283 which at first sight might make us seem a little demented. In our definition of the trigonometric functions we measured the position of C by the angle in degrees which we had turned through, but we could just as well measure the position by the *distance* we've travelled through (one full revolution of that circle of unit radius being 2π in distance). In the following example we have completed one third of a full revolution:

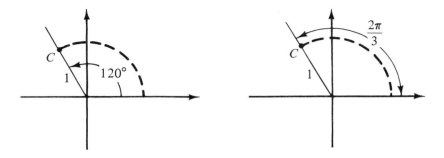

If we measure angles by this distance travelled around the unit circle then the angle is said to be measured in *radians*. Hence 2π (≈ 6.283) radians represent one full revolution and are equivalent to $360°$. This new measure pays off handsomely in our analytical development, as the next example begins to illustrate.

Example	***Example***
(With angle x in **degrees**)	(With angle x in **radians**)

(i) Find the area of the sector and the length of the arc of a circle in terms of the radius r and angle x.

(ii) Show that

$$\lim_{x \to 0} \sin x = 0 \text{ and } \lim_{x \to 0} \cos x = 1$$

(iii) Find

$$\lim_{x \to 0} \frac{\sin x}{x}$$

and interpret this in terms of a gradient of the sine function.

(i) Find the area of the sector and the length of the arc of a circle in terms of the radius r and angle x.

(ii) Show that

$$\lim_{x \to 0} \sin x = 0 \text{ and } \lim_{x \to 0} \cos x = 1$$

(iii) Find

$$\lim_{x \to 0} \frac{\sin x}{x}$$

and interpret this in terms of a gradient of the sine function.

Solution

(i)

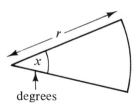

degrees

The area and the circumference of the whole circle are πr^2 and $2\pi r$. Here we have $x/360$ of the circle, so its area and arc length are

$$\frac{x\pi r^2}{360} \text{ and } \frac{x\pi r}{180}$$

(ii)

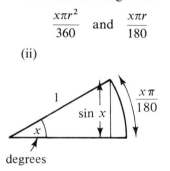

degrees

Solution

(i)

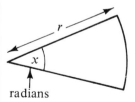

radians

The area and the circumference of the whole circle are πr^2 and $2\pi r$. Here we have $x/2\pi$ of the circle, so its area and arc length are

$$\frac{x r^2}{2} \text{ and } xr$$

(ii)

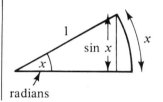

radians

For the acute-angled x as shown it is clear from the figure that

$$0 \leqslant \sin x \leqslant \frac{x\pi}{180}$$

(and for small and negative x

$$-\frac{x\pi}{180} \leqslant \sin x \leqslant 0)$$

Hence

if $x_1, x_2, x_3, \ldots \to 0$

then eventually

$$-\left|\frac{x_n\pi}{180}\right| \leqslant \sin x_n \leqslant \left|\frac{x_n\pi}{180}\right|$$

$$\downarrow \qquad \downarrow \qquad \downarrow$$

$$0 \quad \therefore \quad \downarrow \qquad 0$$

$$0$$

and so $\lim_{x\to 0} \sin x = 0$.

Since $\cos^2 x + \sin^2 x = 1$ and since $\cos x$ is non-negative for small x we have, for such x,

$$\cos x = \sqrt{(1 - \sin^2 x)} \to 1$$

as $x \to 0$; i.e. $\lim_{x\to 0} \cos x = 1$.

Since $\sin 0 = 0$ and $\cos 0 = 1$ these limits show that the sine and cosine functions are continuous at 0.

(iii) For small x

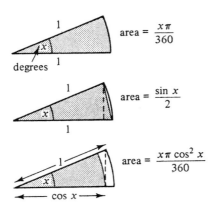

For acute-angled x as shown it is clear from the figure that

$$0 \leqslant \sin x \leqslant x$$

(and for small and negative x

$$-x \leqslant \sin x \leqslant 0)$$

Hence

if $x_1, x_2, x_3, \ldots \to 0$

then eventually

$$-|x_n| \leqslant \sin x_n \leqslant |x_n|$$

$$\downarrow \qquad \downarrow \qquad \downarrow$$

$$0 \quad \therefore \quad \downarrow \qquad 0$$

$$0$$

and so $\lim_{x\to 0} \sin x = 0$.

Since $\cos^2 x + \sin^2 x = 1$ and since $\cos x$ is non-negative for small x we have, for such x,

$$\cos x = \sqrt{(1 - \sin^2 x)} \to 1$$

as $x \to 0$; i.e. $\lim_{x\to 0} \cos x = 1$.

Since $\sin 0 = 0$ and $\cos 0 = 1$ these limits show that the sine and cosine functions are continuous at 0.

(iii) For small x

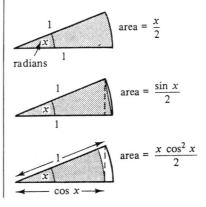

These areas are in decreasing order and so

$$\frac{x\pi \cos^2 x}{360} \leqslant \frac{\sin x}{2} \leqslant \frac{x\pi}{360}$$

and for small positive x we have

$$\frac{\pi \cos^2 x}{180} \leqslant \frac{\sin x}{x} \leqslant \frac{\pi}{180}$$

(and the same inequalities are easily deduced for small $x < 0$). Hence

$$\text{if} \quad \underbrace{x_1, x_2, x_3, \ldots}_{\neq 0} \to 0$$

then eventually

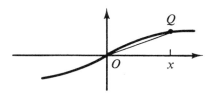

and (when x is in degrees)

$$\lim_{x \to 0} \frac{\sin x}{x} = \frac{\pi}{180}$$

If O is the origin $(0, 0)$ and Q is any other point $(x, \sin x)$ on the graph of the sine function then the gradient of the chord OQ is $\sin x/x$.

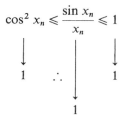

As $x \to 0$ this gradient tends to the gradient of the curve 'at O'. Hence (when x is in degrees) the gradient

---

These areas are in decreasing order and so

$$\frac{x \cos^2 x}{2} \leqslant \frac{\sin x}{2} \leqslant \frac{x}{2}$$

and for small positive x we have

$$\cos^2 x \leqslant \frac{\sin x}{x} \leqslant 1$$

(and the same inequalities are easily deduced for small $x < 0$). Hence

$$\text{if} \quad \underbrace{x_1, x_2, x_3, \ldots}_{\neq 0} \to 0$$

then eventually

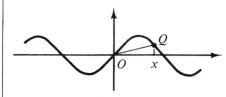

and (when x is in radians)

$$\lim_{x \to 0} \frac{\sin x}{x} = 1$$

If O is the origin $(0, 0)$ and Q is any other point $(x, \sin x)$ on the graph of the sine function then the gradient of the chord OQ is $\sin x/x$.

As $x \to 0$ this gradient tends to the gradient of the curve 'at O'. Hence (where x is in radians) the gradient at $x = 0$ of the sine function is $\pi/180$:

at $x = 0$ of the sine function is 1:

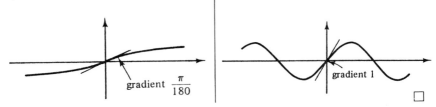

You will have noticed that the answers in the right-hand column are marginally neater than those on the left. But much more important for our subsequent development is that the gradient of the sine function at $x = 0$ should be $\cos 0 \; (=1)$: that works on the right (when x is in radians) but not on the left (when x is in degrees). It is this analytical link between the sine and cosine which makes radians essential: so from this point onwards **our angles will always be measured in radians**. The graphs of the three trigonometric functions are now illustrated. Since $\tan = \sin/\cos$ its graph is easily calculated from the other two and its domain consists of those numbers x for which $\cos x \neq 0$.

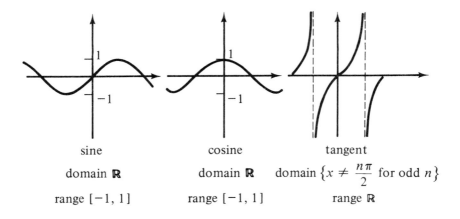

sine

domain $\mathbb{R}$

range $[-1, 1]$

cosine

domain $\mathbb{R}$

range $[-1, 1]$

tangent

domain $\{x \neq \dfrac{n\pi}{2} \text{ for odd } n\}$

range $\mathbb{R}$

Theorem (i) The sine function is continuous.

(ii) For each $x_0 \in \mathbb{R}$

$$\lim_{x \to x_0} \frac{\sin x - \sin x_0}{x - x_0} = \cos x_0$$

Proof Note before proceeding that we can use the formulae for the sine of a sum and a difference of angles to deduce that

$$\sin x - \sin x_0 = \sin\left(\frac{x + x_0}{2} + \frac{x - x_0}{2}\right) - \sin\left(\frac{x + x_0}{2} - \frac{x - x_0}{2}\right)$$

$$= \left(\sin\frac{x + x_0}{2}\cos\frac{x - x_0}{2} + \sin\frac{x - x_0}{2}\cos\frac{x + x_0}{2}\right)$$

$$- \left(\sin\frac{x + x_0}{2}\cos\frac{x - x_0}{2} - \sin\frac{x - x_0}{2}\cos\frac{x + x_0}{2}\right)$$

$$= 2\sin\frac{x - x_0}{2}\cos\frac{x + x_0}{2}$$

(In my day that was one of the trigonometric formulae which we knew by heart!) We can now proceed with the proof of the theorem.

(i) **if** $x_1,$ $x_2,$ $x_3,$ $\ldots \to x_0$

   then $\dfrac{x_1 - x_0}{2},$ $\dfrac{x_2 - x_0}{2},$ $\dfrac{x_3 - x_0}{2},$ $\ldots \to 0$

and (by the continuity of sine at 0 established in the previous example)

$$\sin\frac{x_1 - x_0}{2}, \quad \sin\frac{x_2 - x_0}{2}, \quad \sin\frac{x_3 - x_0}{2}, \quad \ldots \to 0$$

Hence (by the trigonometric identity established above)

$$\sin x_n - \sin x_0 = 2\underbrace{\sin\frac{x_n - x_0}{2}}_{\substack{\downarrow\\0}}\underbrace{\cos\frac{x_n + x_0}{2}}_{\substack{\text{bounded between}\\-1\text{ and }1}}$$

$$\therefore \quad \downarrow \\ 0$$

and so

   then $\sin x_1,$ $\sin x_2,$ $\sin x_3,$ $\ldots \to \sin x_0$

This shows that the sine function is continuous at any x_0, and hence is continuous. (It is easy to deduce that cosine is also continuous, and that is left as one of the next exercises.)

(ii)   **If** $x_1,$ $x_2,$ $x_3,$ $\ldots \to x_0$

                        $\neq x_0$

   then $\dfrac{x_1 - x_0}{2},$ $\dfrac{x_2 - x_0}{2},$ $\dfrac{x_3 - x_0}{2},$ $\ldots \to 0$

   and $\dfrac{x_1 + x_0}{2},$ $\dfrac{x_2 + x_0}{2},$ $\dfrac{x_3 + x_0}{2},$ $\ldots \to x_0$

Therefore (by (iii) of the above example where the angles are measured in radians)

$$\sin \dfrac{x_1 - x_0}{2} \over \dfrac{x_1 - x_0}{2}, \quad \sin \dfrac{x_2 - x_0}{2} \over \dfrac{x_2 - x_0}{2}, \quad \sin \dfrac{x_3 - x_0}{2} \over \dfrac{x_3 - x_0}{2}, \quad \ldots \to 1$$

and (by the continuity of the cosine function)

$$\cos \dfrac{x_1 + x_0}{2}, \quad \cos \dfrac{x_2 + x_0}{2}, \quad \cos \dfrac{x_3 + x_0}{2}, \quad \ldots \to \cos x_0$$

Again using the trigonometric identity above and multiplying these last two convergent sequences together gives

then $$\dfrac{\sin x_n - \sin x_0}{x_n - x_0} = \dfrac{2 \sin \dfrac{x_n - x_0}{2} \cos \dfrac{x_n + x_0}{2}}{x_n - x_0}$$

$$= \underbrace{\dfrac{\sin \dfrac{x_n - x_0}{2}}{\dfrac{x_n - x_0}{2}}}_{\downarrow \atop 1} \times \underbrace{\cos \dfrac{x_n + x_0}{2}}_{\downarrow \atop \cos x_0} \to \cos x_0$$

as required.

We have now interpreted such limits in terms of gradients several times, even though we haven't yet defined differentiation. In this case if P is the point $(x_0, \sin x_0)$ on the graph of the sine function and Q is any other point $(x, \sin x)$ on that graph then

$$\text{gradient of } PQ = \dfrac{\sin x - \sin x_0}{x - x_0}$$

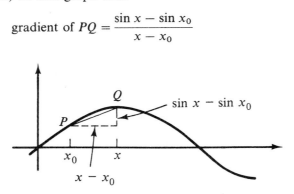

So the gradient of the sine function 'at P' is the limit of that expression as Q approaches P; i.e.

$$\lim_{x \to x_0} \frac{\sin x - \sin x_0}{x - x_0}$$

which we now know is $\cos x_0$.   □

So it seems that the gradient of the sine function at x_0 is $\cos x_0$ (provided of course that we're now measuring our angles in radians). We'll make much more of this in the next chapter.

The sine function is clearly not one-to-one because each number in its range is equal to $\sin x$ for infinitely many values of x; e.g.

$$\tfrac{1}{2} = \sin \frac{\pi}{6} = \sin \frac{5\pi}{6} = \sin \frac{13\pi}{6} = \cdots$$

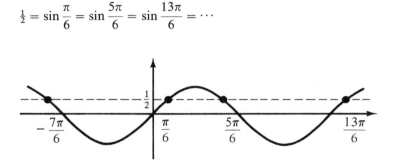

But when a function is not one-to-one (like the function $f(x) = x^2$) its domain can sometimes be restricted to give a function which *is* one-to-one and which therefore has a sensible inverse. (In the case of $f(x) = x^2$ we restricted the domain to the non-negative numbers and obtained the inverse function $f^{-1}(x) = \sqrt{x}$.) In the case of the sine function we'll restrict the domain to the interval $[-\pi/2, \pi/2]$ as shown:

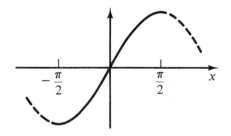

Now with this restricted domain sine is strictly increasing, its domain is the interval $[-\pi/2, \pi/2]$ and its range is $[-1, 1]$. So by the theorem on page 101 it has a continuous inverse function $\sin^{-1}$ with domain $[-1, 1]$ and range $[\pi/2, \pi/2]$. Hence for $x \in [-1, 1]$

$$\sin^{-1} x = y \quad \Leftrightarrow \quad x = \sin y \text{ and } y \in [-\pi/2, \pi/2]$$

and $\sin^{-1} x$ is the unique y chosen in $[-\pi/2, \pi/2]$ with $\sin y = x$.

In a similar way the domain of the cosine function can be restricted to $[0, \pi]$ and the domain of the tangent function to $(-\pi/2, \pi/2)$ to give one-to-one functions with well-defined inverses. Their graphs are illustrated below:

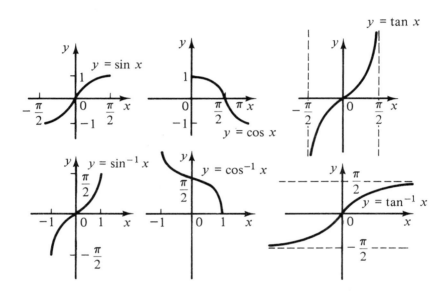

To complete our compendium of the common functions we now define the 'hyperbolic cosine and sine' (and 'hyperbolic tangent' in the exercises): to distinguish these from the trigonometric cosine and sine an 'h' (for hyperbolic) is added.

$$\textit{hyperbolic cosine: } \cosh x = \frac{e^x + e^{-x}}{2} \quad x \in \mathbb{R}$$

$$\textit{hyperbolic sine: } \quad \sinh x = \frac{e^x - e^{-x}}{2} \quad x \in \mathbb{R}$$

There is no real need to define these functions because any application involving them can be achieved (rather less conveniently) in terms of the exponential function. But they do have some interesting properties closely related to those of the cosine and sine functions. The verification of some of these properties will be found in the next exercises and those on gradients will follow from our work on differentiation in the next chapter.

$\cos^2 x + \sin^2 x = 1$	$\cosh^2 x - \sinh^2 x = 1$
For any θ the point $(\cos \theta, \sin \theta)$ lies on the circle $x^2 + y^2 = 1$, and these functions are often referred to as the 'circular' functions.	For any θ the point $(\cosh \theta, \sinh \theta)$ lies on the hyperbola $x^2 - y^2 = 1$, which accounts for their name as the 'hyperbolic' functions.
$\sin(x + y) = \sin x \cos y$ $\qquad\qquad + \sin y \cos x$	$\sinh(x + y) = \sinh x \cosh y$ $\qquad\qquad + \sinh y \cosh x$
$\cos(x + y) = \cos x \cos y$ $\qquad\qquad - \sin y \sin y$	$\cosh(x + y) = \cosh x \cosh y$ $\qquad\qquad + \sinh x \sinh y$
etc.	etc.
Gradient of sin at x_0 is $\cos x_0$.	Gradient of sinh at x_0 is $\cosh x_0$.
Gradient of cos at x_0 is $-\sin x_0$.	Gradient of cosh at x_0 is $\sinh x_0$.

The graphs of the hyperbolic functions are shown below. Since the exponential function is continuous it follows that these hyperbolic functions are also continuous. In fact sinh is strictly increasing with domain and range equal to $\mathbb{R}$ and so it has a well-defined continuous inverse function $\sinh^{-1}$ also with domain and range equal to $\mathbb{R}$. The function cosh is not one-to-one but by restricting its domain to the non-negative numbers it too has an inverse.

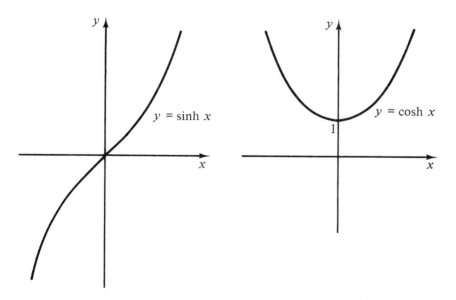

That completes our introductory look at the standard elementary functions of analysis. After some exercises we shall look at some very important properties of these functions (and indeed of *any* continuous functions).

Exercises

1   Show that the cosine function is continuous and that

$$\lim_{x \to x_0} \frac{\cos x - \cos x_0}{x - x_0} = -\sin x_0$$

Interpret that result in terms of gradients.

2   Prove that

$$\sin^{-1} x = \frac{\pi}{2} - \cos^{-1} x$$

for each $x \in [-1, 1]$.

3   Prove that

(i) $\cosh^2 x - \sinh^2 x = 1$
(ii) $\sinh(x + y) = \sinh x \cosh y + \sinh y \cosh x$

and that if tanh is defined by

$$\tanh x = \frac{\sinh x}{\cosh x} \quad x \in \mathbb{R}$$

then

(iii) $\tanh(2x) = \dfrac{2 \tanh x}{1 + \tanh^2 x}$

4   For any $x \in \mathbb{R}$ let $y = \sinh^{-1} x$. Show that

$$e^{2y} - 2xe^y - 1 = 0$$

and deduce that

$$\sinh^{-1} x = \log(x + \sqrt{(1 + x^2)})$$

5   Let the functions f and g be defined by

$$f(x) = \begin{cases} \sin(1/x) & \text{if } x \neq 0 \\ 0 & \text{if } x = 0 \end{cases} \qquad g(x) = \begin{cases} x \sin(1/x) & \text{if } x \neq 0 \\ 0 & \text{if } x = 0 \end{cases}$$

Show that f is not continuous at $x = 0$ but that g is continuous. Sketch the graphs of f and g.

(Solutions on page 257)

Two acts with an interval

As a curtain-raiser let us recall our earlier method for obtaining approximations to roots of an equation.

> **Example**   Let the function f be given by
> $$f(x) = \frac{x^3 + x^2 + 1}{5}$$
> Define a sequence by
> $$x_1 = 0 \quad \text{and} \quad x_n = f(x_{n-1}) \quad \text{for } n > 1$$
> Calculate the first few terms of the sequence and hence obtain, correct to three decimal places, a root of the equation
> $$x^3 + x^2 - 5x + 1 = 0$$
>
> **Solution** $x_1 = 0$
> $$\begin{aligned} x_2 &= f(x_1) = 0.2 \\ x_3 &= f(x_2) = 0.2096 \\ x_4 &= f(x_3) \approx 0.210\,63 \\ x_5 &= f(x_4) \approx 0.210\,74 \\ x_6 &= f(x_5) \approx 0.210\,75 \end{aligned}$$

As we've seen before if this sequence converges to limit x_0 say,
$$x_1, x_2, x_3, \ldots \to x_0$$
then (by the obvious continuity of f)
$$\left. \begin{array}{ccc} f(x_1), f(x_2), f(x_3), \ldots \to f(x_0) \\ \| \qquad \| \qquad \| \\ x_2, \quad x_3, \quad x_4, \ldots \to x_0 \end{array} \right\} \therefore \; x_0 = f(x_0)$$

In this case the sequence seems to be converging to approximately 0.211 and so we hope that this is an approximate solution of the equation
$$x = f(x) = \frac{x^3 + x^2 + 1}{5} \quad \text{or} \quad x^3 + x^2 - 5x + 1 = 0$$

To check that 0.211 *is* a root of that equation correct to three decimal places we'll calculate the values of the cubic at $x = 0.2105$ and $x = 0.2115$:

at $x = 0.2105$ the cubic $x^3 + x^2 - 5x + 1 \approx \quad 0.001 > 0$
at $x = 0.2115$ the cubic $x^3 + x^2 - 5x + 1 \approx -0.003 < 0$

$\left. \begin{array}{l} \text{Hence the cubic} \\ \textit{is } 0 \text{ somewhere} \\ \text{between these} \\ \text{two points (?)} \end{array} \right.$

Therefore there is a root of cubic which, when corrected to three decimal places, equals 0.211.   □

To a pure mathematician that method raises two obvious questions. Firstly, is it a foolproof technique which can be used for finding an approximation to a root of any equation? Sadly not. For instance using the same function f as in that example but with $x_1 = 2$ we get the sequence

$$x_1 = 2, \quad x_2 = 2.6, \quad x_3 = 5.0672, \quad x_4 \approx 31.36, \quad x_5 \approx 6363, \quad x_6 \approx 5 \times 10^{10}$$

which is clearly not converging.

Secondly, how is that last step in the solution justified? Just because the cubic is positive at one point and negative at another how can we be sure that it is zero somewhere between them? I suppose it's because we are familiar with the shape of the graph of a cubic and know that it will behave perfectly reasonably between those two points.

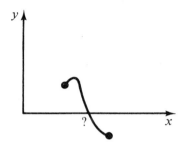

That simple property, namely that if a function is positive at one point and negative at another then it is zero somewhere between them, holds for many functions and it forms the basis of a much more foolproof method for obtaining approximations to solutions of an equation.

> ***Example*** Find a number which is within 0.01 of a root of the equation
>
> $$x^4 - 3x^3 - 2x^2 + 1 = 0$$

Solution Let $f(x) = x^4 - 3x^3 - 2x^2 + 1 \ (x \in \mathbb{R})$. The first step is to find some x with $f(x) < 0$ and another x with $f(x) > 0$. That's easy in this case because clearly $f(1) < 0$ and $f(0) > 0$.

Let

$l_1 = 1$ $\qquad\qquad\qquad\qquad$ $m_1 = 0$

(so that $f(l_1)$ is **Less** than 0) $\qquad$ (so that $f(m_1)$ is **More** than 0)

Then consider the mid-point (or average) of l_1 and m_1, namely 0.5. My calculator gives the value of $f(0.5)$ as 0.1875 which is **More** than 0. So we'll use the 0.5 to replace the m and leave the l unchanged:

$l_2 = 1$ $\qquad\qquad\qquad\qquad$ $m_2 = 0.5$

(so that $f(l_2)$ is **Less** than 0) $\qquad$ (so that $f(m_2)$ is **More** than 0)

Now the mid-point of this l and m is 0.75 and $f(0.75) \approx -1.07$ which is Less than 0. So we'll use the 0.75 to replace the l and leave the m unchanged:

$$l_3 = 0.75 \hspace{3cm} m_3 = 0.5$$

(so that $f(l_3)$ is Less than 0)   (so that $f(m_3)$ is More than 0)

At each stage the root we're looking for is sandwiched between the l_n and the m_n, and since the l_n and m_n are getting closer and closer together we are narrowing down the region in which the root can lie. In fact

$$l_1 - m_1 = 1$$
$$l_2 - m_2 = \tfrac{1}{2}$$
$$l_3 - m_3 = \tfrac{1}{4}$$

with the difference halving each time. If we continue until $l_n - m_n$ is less than 0.02 then as the root is between l_n and m_n it follows that the mid-point of l_n and m_n is less than 0.01 from the root. In fact

$$l_7 - m_7 = \frac{1}{2^6} = \frac{1}{64} < 0.02$$

and so the mid-point of l_7 and m_7 will give us our required approximation. Now completing our calculations we get

f **Less than 0** $\hspace{2cm}$ f **More than 0**

$l_1 = 1 \hspace{4cm} m_1 = 0$

$\hspace{1.5cm} f(0.5) > 0$

$l_2 = 1 \hspace{4cm} m_2 = 0.5$

$\hspace{1.5cm} f(0.75) < 0$

$l_3 = 0.75 \hspace{3.3cm} m_3 = 0.5$

$\hspace{1.5cm} f(0.625) < 0$

$l_4 = 0.625 \hspace{3cm} m_4 = 0.5$

$\hspace{1.5cm} f(0.5625) < 0$

$l_5 = 0.5625 \hspace{2.7cm} m_5 = 0.5$

$\hspace{1.5cm} f(0.53125) > 0$

$l_6 = 0.5625 \hspace{2.7cm} m_6 = 0.53125$

$\hspace{1.5cm} f(0.546875) > 0$

$l_7 = 0.5625 \hspace{2.7cm} m_7 = 0.546875$

and it is clear that 0.554 is within 0.01 of the root. Of course this process can be continued to obtain an approximation to the root to greater and greater degrees of accuracy. Writing a short program to execute the above process gave the root as 0.547 025 09 correct to eight decimal places (and in fact $f(0.547\ 025\ 09) \approx 0.000\ 000\ 000\ 1$, which is pretty close to the required zero).   □

For obvious reasons that technique is known as the 'bisection method'. It is summed up by the following flow-chart which can of course easily be executed by a simple program on a computer. As we shall soon see it works for a wide variety of equations and to any required degree of accuracy $\varepsilon\ (>0)$:

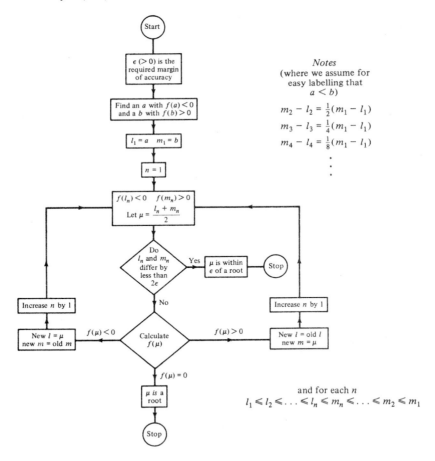

This method can be applied to many equations of the form $f(x) = 0$ but it requires a key property of f:

**if $f(l) < 0$ and $f(m) > 0$ then $f(x) = 0$ for   some x
between l and m**

That is certainly not true of every function as the following examples
illustrate:

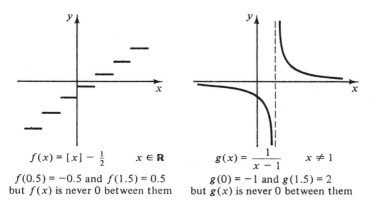

$$f(x) = [x] - \tfrac{1}{2} \qquad x \in \mathbf{R}$$

$f(0.5) = -0.5$ and $f(1.5) = 0.5$
but $f(x)$ is never 0 between them

$$g(x) = \frac{1}{x - 1} \qquad x \ne 1$$

$g(0) = -1$ and $g(1.5) = 2$
but $g(x)$ is never 0 between them

The property failed here because of a break in the graph of each of the
functions. In the case of f that break was because of a discontinuity and in
the case of g that break was because its domain did not include all the
numbers between 0 and 1.5. So to find some functions which satisfy the
above property we are certainly going to restrict attention to continuous
functions whose domains are intervals. The key to success turns out to be
the following theorem:

> **Theorem** Let f be a continuous function whose domain is an
> interval. Then its range is also an interval.

> **Proof** To show that the range of f is an interval we take two
> numbers in that range ($f(a)$ and $f(b)$ say) and show that all numbers
> between them are also in the range. So assume that $f(a) < \gamma < f(b)$: we
> must show that γ is also in f's range; i.e. we must find a number c in f's
> domain with $f(c) = \gamma$.

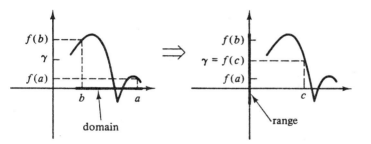

Our proof simply adapts the bisection method described above. Consider the following revised version of our flow-chart for that method:

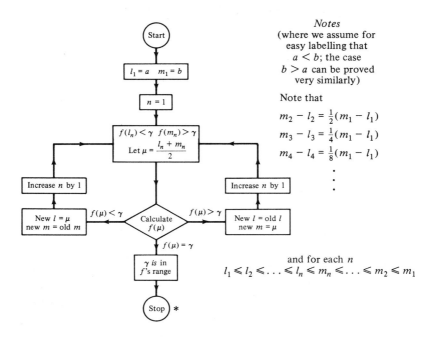

Notes
(where we assume for easy labelling that $a < b$; the case $b > a$ can be proved very similarly)

Note that
$$m_2 - l_2 = \tfrac{1}{2}(m_1 - l_1)$$
$$m_3 - l_3 = \tfrac{1}{4}(m_1 - l_1)$$
$$m_4 - l_4 = \tfrac{1}{8}(m_1 - l_1)$$
$$\vdots$$

and for each n
$$l_1 \leqslant l_2 \leqslant \ldots \leqslant l_n \leqslant m_n \leqslant \ldots \leqslant m_2 \leqslant m_1$$

This process either stops at ∗ when it has found a number μ with $f(\mu) = \gamma$ (thus showing that γ *is in* f's range) or it continues churning out sequences (l_n) and (m_n) with

$$m_n - l_n = \frac{1}{2^{n-1}}(m_1 - l_1)$$

and

$$l_1 \leqslant l_2 \leqslant l_3 \leqslant \cdots \leqslant l_{n-1} \leqslant l_n \leqslant m_n \leqslant m_{n-1} \leqslant \cdots \leqslant m_3 \leqslant m_2 \leqslant m_1$$
$$\underbrace{\hspace{2cm}}_{f(l_n) < \gamma} \quad \underbrace{\hspace{2cm}}_{f(m_n) > \gamma}$$

Hence the sequence (l_n) is increasing and bounded above and so it converges (to c say); and the sequence (m_n) is decreasing and bounded below and so it converges (to c' say). Therefore the sequence $(m_n - l_n)$ converges to $c' - c$. But as $n \to \infty$

$$m_n - l_n = \frac{1}{2^{n-1}}(m_1 - l_1) \to 0$$

and so $c' - c$ is 0 and $c' = c$. This means (as you might have expected from the above construction) that the sequences (l_n) and (m_n) converge to the same limit:

$$l_1, l_2, l_3, \ldots \to c \quad \text{and} \quad m_1, m_2, m_3, \ldots \to c$$

Now by the continuity of f and the fact that a convergent sequence of terms all less than or equal to γ has limit less than or equal to γ, etc, we have

$$f(l_1), f(l_2), f(l_3), \ldots \to f(c) \quad \text{and} \quad f(m_1), f(m_2), f(m_3), \ldots \to f(c)$$
$$<\gamma \quad <\gamma \quad <\gamma \quad \therefore \leqslant\gamma \qquad \qquad >\gamma \quad >\gamma \quad >\gamma \quad \therefore \geqslant\gamma$$

Hence $f(c) = \gamma$ is indeed in f's range.

This shows that f's range is an interval as required. □

> **Corollary** (*The intermediate value theorem*) Let f be a continuous function whose domain includes $[a, b]$. Then if γ is between $f(a)$ and $f(b)$ it follows that $\gamma = f(c)$ for some c between a and b.

> **Solution** Simply apply the theorem to the function f with its domain restricted to the interval $[a, b]$. □

The following graphs illustrate some continuous functions whose domains (and therefore whose ranges) are intervals. The examples include log and sine where until now we couldn't be absolutely certain about their ranges.

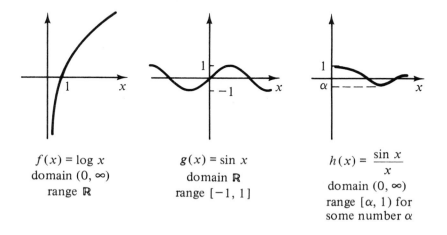

$f(x) = \log x$
domain $(0, \infty)$
range $\mathbb{R}$

$g(x) = \sin x$
domain $\mathbb{R}$
range $[-1, 1]$

$h(x) = \dfrac{\sin x}{x}$
domain $(0, \infty)$
range $[\alpha, 1)$ for some number α

By the *maximum* and *minimum* value of a function we shall mean the maximum and minimum numbers in its range (if they exist). So in the above examples *f* has no maximum or minimum since its range isn't even bounded above or bounded below. The function *g* has a maximum of 1 (achieved for example when $x = \pi/2$ and when $x = 5\pi/2$) and a minimum of -1 (achieved for example when $x = -\pi/2$ and when $x = 3\pi/2$). The function *h* has range $[\alpha, 1)$ for some number α, which will be hard to find until we have studied 'stationary' (or 'turning') points in the next chapter: hence *h* has a minimum of α, but it has no maximum because, as we've seen before, if you give me any number *y* in $[\alpha, 1)$ there is a bigger number (such as $\frac{1}{2}(y + 1)$) still in $[\alpha, 1)$.

We have just seen in the last theorem that if a continuous function *f* has an interval as its domain then its range will also be an interval. It turns out very neatly that if *f*'s domain is a closed and bounded interval (i.e. one of the form $[a, b]$) then so is its range: such an *f* then has a maximum and a minimum value.

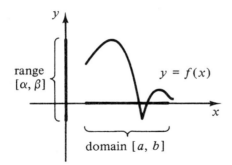

Theorem Let *f* be a continuous function whose domain is a closed and bounded interval. Then its range is also a closed and bounded interval; in particular *f* has a maximum and a minimum.

Proof Assume that *f*'s domain is the non-empty interval $[a, b]$. We know from the previous theorem that its range is also an interval: to show that it's of the form $[\alpha, \beta]$ we must simply show that it has a minimum and a maximum. We shall show that the range has a maximum β, the case of the minimum following very similarly.

We shall need one crucial result from sequences, namely that any sequence in $[a, b]$ has a subsequence convergent to some limit in $[a, b]$ (exercise 4, page 50).

So now assume that f's range is *not* bounded above. Then

there exists $x_1 \in [a, b]$ with $f(x_1) > 1$
and there exists $x_2 \in [a, b]$ with $f(x_2) > 2$
and there exists $x_3 \in [a, b]$ with $f(x_3) > 3$

$$\vdots$$

As we've just commented the sequence (x_n) has a subsequence convergent to some point in $[a, b]$:

$$x_{k_1}, x_{k_2}, x_{k_3}, \ldots \to x_0$$

Note that for each n we have $f(x_{k_n}) > k_n \geqslant n$ and by the continuity of f we have

$$f(x_{k_1}), f(x_{k_2}), f(x_{k_3}), \ldots \to f(x_0)$$
$$> 1 \quad > 2 \quad > 3$$

But then we have a convergent sequence which is also tending to infinity! That contradiction shows that f's range *is* bounded above.

So f's range is non-empty and bounded above: it therefore has a least upper bound β: our aim is to show that β is in the range (and is hence a maximum member of the range). Since β is the *least* upper bound of the range, $\beta - 1$ is not an upper bound of the range, nor is $\beta - \frac{1}{2}$ nor $\beta - \frac{1}{3}$ etc. Hence

there exists $x_1' \in [a, b]$ with $f(x_1') > \beta - 1$
and there exists $x_2' \in [a, b]$ with $f(x_2') > \beta - \frac{1}{2}$
and there exists $x_3' \in [a, b]$ with $f(x_3') > \beta - \frac{1}{3}$

$$\vdots$$

As before the sequence (x_n') has a subsequence convergent to some point in $[a, b]$:

$$x_{k_1}', x_{k_2}', x_{k_3}', \ldots \to x_0'$$

and for each n we have $f(x_{k_n}') > \beta - 1/k_n \geqslant \beta - 1/n$. Hence as $n \to \infty$

$$\beta - 1/n < f(x_{k_n}') \leqslant \beta$$
$$\downarrow \qquad \qquad \downarrow$$
$$\beta \qquad \therefore \downarrow \qquad \beta$$
$$\beta$$

But the sequence $(f(x_{k_n}'))$ converges to $f(x_0')$ since f is continuous. It therefore follows that $\beta = f(x_0')$ and β *is* in f's range.

Hence the supremum (least upper bound) of f's range is actually contained in the range and is the maximum value which the function takes.

As commented earlier the proof for the minimum works in a very similar way and the theorem follows.   □

That completes our study of the continuity of functions. All the standard functions turned out to be continuous and hence all combinations of them formed by adding, subtracting, multiplying, dividing and forming composites of them are still continuous. This gives a wealth of functions to which these latest theorems and techniques apply.

Informally we have already begun to see the uses of gradients and now at last, after these next exercises, we shall be ready to study calculus.

Exercises

1   Let f be a continuous function whose domain includes the interval $[a, b]$ and assume that $f(a) < 0$ and $f(b) > 0$. Suppose that we wish to find a number within ε (>0) of a root of the equation $f(x) = 0$. Show that the bisection method outlined in the flow-chart on page 136 will always terminate at such a number.

Use the bisection method to find a number within 0.002 of a root of the equation $\tan x - x = 0$ between $\pi/2$ and $3\pi/2$.

(In fact that root turns out to be precisely the point where the function $h(x) = (\sin x)/x$ ($x > 0$) achieves its minimum, as illustrated on page 139.)

2   Show that a cubic equation (i.e. one of the form $ax^3 + bx^2 + cx + d = 0$ where $a \neq 0$) has at least one real root.

3   Let f be a continuous function whose domain is $[a, b]$ and whose range is a subset of $[a, b]$. By considering the function g given by $g(x) = f(x) - x$ show that there is at least one x_0 in $[a, b]$ with $x_0 = f(x_0)$.

Consider the function f given by $f(x) = \cos x$ ($x \in [0, \pi/2]$). Show that the range is a subset of $[0, \pi/2]$ and by sketching the graphs of $y = f(x)$ and of $y = x$ observe that in this case there is precisely one x_0 with $x_0 = f(x_0)$.

Choose any $x_1 \in [0, \pi/2]$ and (as in exercise 3 on page 107) define a sequence (x_n) by $x_n = f(x_{n-1})$ for each $n > 1$. Calculate the first few terms of this sequence and hence find, correct to four decimal places, the unique root of the equation $\cos x - x = 0$.

4   Let f be a continuous function with domain $[a, b]$. By the last theorem f's range is of the form $[\alpha, \beta]$ for some α and β. Assume that $f(x_0) = \beta$ and let P be the point $(x_0, f(x_0))$ on the graph of f. Show that

(i) if $x_0 < x \leq b$ and Q is the point $(x, f(x))$ on the graph of f, then the gradient of the chord PQ is less than or equal to 0;

(ii) if $a \leq x < x_0$ and Q is the point $(x, f(x))$ on the graph of f, then the gradient of the chord PQ is greater than or equal to 0.

(Solutions on page 259)

4

Calculus at last

Differentiation

We have already discussed gradients several times in the previous chapters. To find the gradient of a function f at the point P on its graph we look at a nearby point Q on the graph (to the left or right of P) and calculate the gradient of the chord PQ. We then consider what happens to that gradient as Q approaches P:

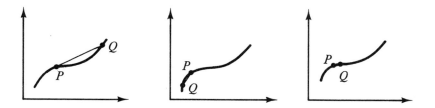

If P is the point $(x_0, f(x_0))$ and Q is $(x, f(x))$ then the gradient of the chord PQ is

$$\frac{f(x) - f(x_0)}{x - x_0}$$

As we introduced in our work on limits, the value (if any) which this approaches as Q approaches P is denoted by

$$\lim_{x \to x_0} \frac{f(x) - f(x_0)}{x - x_0}$$

As you will know from school, differentiation is closely connected with the idea of a gradient, and we can now formally introduce the concept of a function f being 'differentiable' at x_0:

Definition   The function f is *differentiable at the point* x_0 in its domain if

$$\lim_{x \to x_0} \frac{f(x) - f(x_0)}{x - x_0}$$

exists. In that case the limit is denoted by $f'(x_0)$ (or df/dx – but more about that notation later) and is called the *derivative* of f at x_0. If a function is differentiable at each point of its domain then it is simply called *differentiable*.

So that we can consider the limit in the definition in a meaningful way there must be at least one sequence (x_n) which converges to x_0 for which

$$\frac{f(x_n) - f(x_0)}{x_n - x_0}$$

can be calculated; i.e. a sequence (x_n) which converges to x_0, which is in f's domain and none of whose terms equals x_0. We shall only ever consider the differentiability of a function at such points. When the domain of f includes a whole interval around x_0 we can actually visualise the gradient of f at x_0: if a sensible gradient exists its value will be $f'(x_0)$. As you probably already know, the process of finding the derivative is known as *differentiation*.

Example   Let f be the function given by

$$f(x) = x^2 \quad x \in \mathbb{R}$$

Show that f is differentiable and that $f'(x_0) = 2x_0$ for each x_0. Draw the graph of f and observe that its gradient at $x = 1$ does indeed seem to be 2.

Solution   The function f is differentiable at x_0 if

$$\lim_{x \to x_0} \frac{f(x) - f(x_0)}{x - x_0}$$

exists. But that limit is

$$\lim_{x \to x_0} \frac{x^2 - x_0{}^2}{x - x_0} = \lim_{x \to x_0} (x + x_0) = 2x_0$$

(Are you happy to accept that $\lim_{x \to x_0}(x + x_0) = 2x_0$? You could use the fact that sums of limits behave in a nice way and note that

$$\lim_{x \to x_0} (x + x_0) = \lim_{x \to x_0} x + \lim_{x \to x_0} x_0 = x_0 + x_0 = 2x_0$$

Alternatively you could go back to basics and note that

> **if** x_1, x_2, x_3, $\ldots \to x_0$
>
> **then** $x_1 + x_0$, $x_2 + x_0$, $x_3 + x_0$, $\ldots \to 2x_0$

We shall gradually dispense with these formal verifications.)

Hence for each x_0 we have

$$f'(x_0) = \lim_{x \to x_0} \frac{x^2 - x_0^2}{x - x_0} = 2x_0$$

as required.

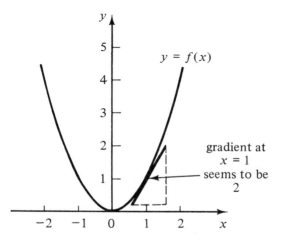

A reasonably accurate drawing of the graph of f shows that the gradient of f at $x = 1$ does seem to be 2 as expected. $\square$

If a function f has domain D then at each point x_0 in D at which f is differentiable there is a well-defined derivative $f'(x_0)$. So f' is another function, called the *derivative* of f, whose domain is a subset of D.

> ***Example***   Let f be the function given by
>
> $$f(x) = |x| \quad x \in \mathbb{R}$$
>
> Show that f is differentiable at each non-zero point but is not differentiable at $x = 0$. Show that its derivative is given by
>
> $$f'(x) = \begin{cases} 1 & \text{if } x > 0 \\ -1 & \text{if } x < 0 \end{cases}$$

Solution Let $x_0 > 0$. Then we shall show that

$$\lim_{x \to x_0} \frac{f(x) - f(x_0)}{x - x_0} = 1$$

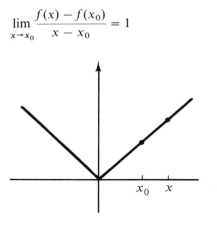

If $x_1, x_2, x_3, \ldots \to x_0$

$\underbrace{\qquad\qquad}_{\neq x_0} \qquad > 0$

then eventually the terms are all positive (in which case $f(x_n) = x_n$) and so eventually

then $\dfrac{f(x_n) - f(x_0)}{x_n - x_0} = \dfrac{x_n - x_0}{x_n - x_0} \to 1$

and so for $x_0 > 0$ we have

$$f'(x_0) = \lim_{x \to x_0} \frac{f(x_n) - f(x_0)}{x_n - x_0} = 1$$

Similarly

    if $x_1, x_2, x_3, \ldots \to x_0$

$\underbrace{\qquad\qquad}_{\neq x_0} \qquad < 0$

then eventually the terms are all negative (in which case $f(x_n) = -x_n$) and so eventually

then $\dfrac{f(x_n) - f(x_0)}{x_n - x_0} = \dfrac{(-x_n) - (-x_0)}{x_n - x_0} \to -1$

Hence for $x_0 < 0$ we have

$$f'(x_0) = \lim_{x \to x_0} \frac{f(x_n) - f(x_0)}{x_n - x_0} = -1$$

Finally we consider the differentiability of f at 0 itself: as you can see from the graph or from our work above, to the right of 0 the gradient is 1 and to the left of 0 the gradient is -1. So to confirm formally that f is not differentiable at 0 we consider the sequence

$$1, -\tfrac{1}{2}, \tfrac{1}{3}, -\tfrac{1}{4}, \ldots \to 0$$

But then the sequence

$$\frac{f(1) - f(0)}{1 - 0}, \quad \frac{f(-\tfrac{1}{2}) - f(0)}{-\tfrac{1}{2} - 0}, \quad \frac{f(\tfrac{1}{3}) - f(0)}{\tfrac{1}{3} - 0}, \quad \frac{f(-\tfrac{1}{4}) - f(0)}{-\tfrac{1}{4} - 0}, \ldots$$

$$\| \qquad\qquad \| \qquad\qquad \| \qquad\qquad \|$$

$$1 \qquad\qquad -1 \qquad\qquad 1 \qquad\qquad -1$$

does not converge. Hence

$$\lim_{x \to 0} \frac{f(x) - f(0)}{x - 0}$$

does not exist and f is not differentiable at $x = 0$.

Putting the three parts of our solution together we deduce that the domain of the derivative f' is the set of non-zero numbers and that f' is given by

$$f'(x) = \begin{cases} 1 & \text{if } x > 0 \\ -1 & \text{if } x < 0 \end{cases}$$

as required.   □

We have remarked that if f is a function with domain D then f' is a function whose domain is a subset of D. Actually if the behaviour of f is perverse enough then it may fail to be differentiable anywhere (giving f' an empty domain!).

Example   Let f be the function given by

$$f(x) = \begin{cases} 0 & \text{if } x \text{ is rational} \\ 1 & \text{if } x \text{ is irrational} \end{cases}$$

Show that f is differentiable nowhere.

Solution   We deal firstly with a rational x_0. There exists a sequence of numbers different from x_0 which converges to x_0 and such that the odd terms are rational and the even terms irrational; e.g.

$$x_1, \qquad x_2, \qquad x_3, \qquad x_4, \qquad \ldots$$

$$\| \qquad\quad \| \qquad\quad \| \qquad\quad \|$$

$$x_0 + 1, \quad x_0 + \frac{\sqrt{2}}{2}, \quad x_0 + \frac{1}{3}, \quad x_0 + \frac{\sqrt{2}}{4}, \quad \ldots$$

But then

$$\frac{f(x_n) - f(x_0)}{x_n - x_0} = \begin{cases} \dfrac{0 - 0}{1/n} = 0 & \text{if } n \text{ is odd} \\[2ex] \dfrac{1 - 0}{\sqrt{2}/n} = \dfrac{n}{\sqrt{2}} & \text{if } n \text{ is even} \end{cases}$$

and so the sequence

$$\underset{\substack{\| \\ 0}}{\frac{f(x_1) - f(x_0)}{x_1 - x_0}}, \qquad \underset{\substack{\| \\ \frac{2}{\sqrt{2}}}}{\frac{f(x_2) - f(x_0)}{x_2 - x_0}}, \qquad \underset{\substack{\| \\ 0}}{\frac{f(x_3) - f(x_0)}{x_3 - x_0}},$$

$$\underset{\substack{\| \\ \frac{4}{\sqrt{2}}}}{\frac{f(x_4) - f(x_0)}{x_4 - x_0}}, \qquad \ldots$$

clearly does not converge. Hence

$$\lim_{x \to x_0} \frac{f(x) - f(x_0)}{x - x_0}$$

does not exist and f is not differentiable at x_0.

A very similar argument works for an irrational x_0 (although it's not so easy to be specific about the numerical values of the sequence (x_n)). $\qquad \square$

It is not surprising that the function in that example failed to be differentiable because it is not even continuous. Informally continuity means that the function has no unnatural break, and differentiability means that the function has a gradient or is 'smooth'. So one would perhaps expect a differentiable function to be continuous, which we now prove.

> **Theorem**   If a function is differentiable at a point then it is continuous there.

> **Proof**   Let the function f be differentiable at x_0: we must show that it is continuous there. So note that

$$\textbf{if} \quad \underbrace{x_1, x_2, x_3, \ldots}_{\textit{in } f\text{'s domain and } \neq x_0} \to x_0$$

then, since f is differentiable at x_0, as $n \to \infty$ we have

$$\frac{f(x_n) - f(x_0)}{x_n - x_0} \to f'(x_0)$$

But then as convergent sequences can be sensibly multiplied together we deduce that

$$f(x_n) - f(x_0) = \frac{f(x_n) - f(x_0)}{x_n - x_0} \times (x_n - x_0) \to f'(x_0) \times 0 = 0$$

and so

$$\textbf{then} \quad f(x_1), \quad f(x_2), \quad f(x_3), \dots \to f(x_0)$$

It therefore follows that f is continuous at x_0, and the theorem is proved.

□

We saw in the first example that if $f(x) = x^2$ ($x \in \mathbb{R}$) then $f'(x) = 2x$ ($x \in \mathbb{R}$): in the exercises and in the next section we shall extend that idea to other powers and we shall gradually build up a list of standard derivatives. This list already includes some results which follow easily from the work we did on limits in chapter 3 (see the table): in each of these cases the derivative has the same domain as the function itself. After the next exercises we shall gradually be able to extend this list by learning how to differentiate various combinations of differentiable functions.

$f(x)$	$f'(x)$	
x^2	$2x$	
x^n	nx^{n-1}	(exercises)
e^x	e^x	(page 110)
$\log x$	$1/x$	(page 118)
$\sin x$	$\cos x$	(page 126)
$\cos x$	$-\sin x$	(page 132)

Exercises

1   Let m be a positive integer. Show that for any numbers x and x_0

$$(x - x_0)(x^{m-1} + x^{m-2}x_0 + x^{m-3}x_0^2 + \cdots + xx_0^{m-2} + x_0^{m-1})$$
$$= x^m - x_0^m$$

and deduce that

$$\lim_{x \to x_0} \frac{x^m - x_0^m}{x - x_0} = mx_0^{m-1}$$

Hence show that if the function f is given by $f(x) = x^m$ $(x \in \mathbb{R})$ then f is differentiable with derivative given by $f'(x) = mx^{m-1}$.

2   Let f be a differentiable function with domain D, let k be any number and let g be the function given by

$$g(x) = kf(x) \quad x \in D$$

Show that g is differentiable and that $g'(x) = kf(x)$ for each $x \in D$.

3   (i) Let f be a function which is differentiable at x_0 with $f(x_0) \neq 0$ and let g be the function given by $g(x) = 1/f(x)$. Show that

$$\lim_{x \to x_0} \frac{g(x) - g(x_0)}{x - x_0} = \lim_{x \to x_0} -\frac{1}{f(x)} \cdot \frac{1}{f(x_0)} \cdot \frac{f(x) - f(x_0)}{x - x_0}$$

and deduce that g is differentiable at x_0 with derivative

$$g'(x_0) = -\frac{f'(x_0)}{(f(x_0))^2}$$

(ii) Now let $g(x) = x^{-m}$ $(x \neq 0)$ where m is a positive integer. Use the first part of this exercise to show that $g'(x) = -mx^{-m-1}$.

Hence show that the rule

$$f(x) = x^n \quad \Rightarrow \quad f'(x) = nx^{n-1}$$

works for any integer n.

4   (i) Let l be a line of gradient α $(\neq 0)$ in the coordinate (x, y)-plane and let l' be the reflection of l in the line $y = x$: find the gradient of l'.

Now let f be a function with inverse g. If x_0 is in f's domain and g has gradient α $(\neq 0)$ at $f(x_0)$, as illustrated below, what do you expect the gradient of f to be at x_0?

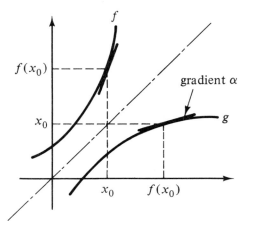

Now prove that result formally: assume that g is differentiable at $f(x_0)$ with $g'(f(x_0)) = \alpha$ $(\neq 0)$ and deduce from the definition of the derivative that $f'(x_0) = 1/\alpha$.

(ii) Now let m be a positive integer and let f and g be the functions given by

$$f(x) = x^{1/m} \quad (x > 0) \quad \text{and} \quad g(x) = x^m \quad (x > 0)$$

Use the first part of the exercise to show that $f'(x) = (1/m)x^{(1/m) - 1}$.

(We shall extend this rule even further in the next exercises and show that

$$f(x) = x^r \Rightarrow f'(x) = rx^{r-1}$$

for any rational number r, thus confirming that differentiation of powers of x is exactly as expected.)

(Solutions on page 262)

Combinations of functions

As we have seen before with other analytical concepts, once we have applied the definition in a few cases we learn some short-cuts and soon leave the formal definition behind. For example consider the function given by

$$f(x) = x^2 + x^3 \quad x \in \mathbb{R}$$

Its gradient at $x = 1$ for example turns out to be the sum of the gradients of $f_1(x) = x^2$ and $f_2(x) = x^3$ at $x = 1$:

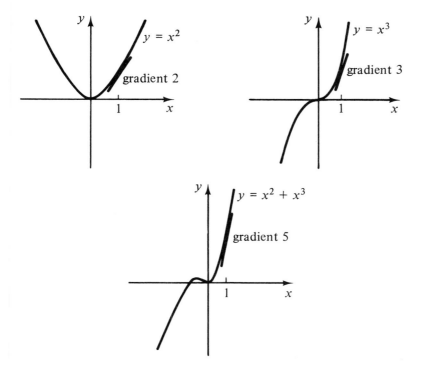

This is one of those properties of differentiation which you have always taken for granted. We are now able to derive analytically all the usual properties concerning the derivatives of sums, differences, products and quotients of functions. As usual when we talk about $f + g$ (for example) its domain is the intersection of the domains of f and g, and when we talk about the differentiability of a combination of functions we assume that their two domains are suitably overlapping.

> ***Theorem*** Let the functions f and g be differentiable at x_0. Then
>
> (i) the function $f + g$ is differentiable at x_0 with
>
> $$(f + g)'(x_0) = f'(x_0) + g'(x_0)$$
>
> (ii) the function $f - g$ is differentiable at x_0 with
>
> $$(f - g)'(x_0) = f'(x_0) - g'(x_0)$$
>
> (iii) the function fg (the product of f and g) is differentiable at x_0 with
>
> $$(fg)'(x_0) = f(x_0)g'(x_0) + f'(x_0)g(x_0)$$
>
> (iv) provided that $g(x_0) \neq 0$ the function f/g is differentiable at x_0 with
>
> $$\left(\frac{f}{g}\right)'(x_0) = \frac{g(x_0)f'(x_0) - g'(x_0)f(x_0)}{(g(x_0))^2}$$

Proof We are given that

if $\underbrace{x_1, \quad x_2, \quad x_3, \quad \ldots}_{\textit{in } f\text{'s and } g\text{'s domain, } \neq x_0} \to x_0$

then $\dfrac{f(x_n) - f(x_0)}{x_n - x_0} \to f'(x_0)$ and $\dfrac{g(x_n) - g(x_0)}{x_n - x_0} \to g'(x_0)$

as $n \to \infty$

(i) Therefore

if $\underbrace{x_1, \quad x_2, \quad x_3, \quad \ldots}_{\textit{in } f\text{'s and } g\text{'s domain, } \neq x_0} \to x_0$

then $\dfrac{(f + g)(x_n) - (f + g)(x_0)}{x_n - x_0} = \dfrac{f(x_n) + g(x_n) - f(x_0) - g(x_0)}{x_n - x_0}$

$= \dfrac{f(x_n) - g(x_0)}{x_n - x_0} + \dfrac{g(x_n) - g(x_0)}{x_n - x_0} \to f'(x_0) + g'(x_0)$

as $n \to \infty$ and so $f + g$ is differentiable at x_0 and

$$(f + g)'(x_0) = \lim_{x \to x_0} \frac{(f + g)(x) - (f + g)(x_0)}{x - x_0} = f'(x_0) + g'(x_0)$$

as required.

(ii) Either a very similar argument to that in (i), or the result of (i) applied to the functions f and $(-1) \times g$ gives the required result concerning $f - g$.

(iii) Similarly

if $x_1, \quad x_2, \quad x_3, \quad \dots \to x_0$

$\underbrace{\qquad\qquad\qquad}_{\text{in } f\text{'s and } g\text{'s domain, } \neq x_0}$

then $\dfrac{(fg)(x_n) - (fg)(x_0)}{x_n - x_0}$

$$= \frac{f(x_n)g(x_n) - f(x_0)g(x_0)}{x_n - x_0}$$

$$= \frac{f(x_n)g(x_n) - f(x_n)g(x_0) + f(x_n)g(x_0) - f(x_0)g(x_0)}{x_n - x_0}$$

$$= f(x_n)\frac{g(x_n) - g(x_0)}{x_n - x_0} + \frac{f(x_n) - f(x_0)}{x_n - x_0}g(x_0)$$

$$\to f(x_0)g'(x_0) + f'(x_0)g(x_0) \quad \text{as } n \to \infty$$

Hence fg is differentiable at x_0 and

$$(fg)'(x_0) = \lim_{x \to x_0} \frac{(fg)(x) - (fg)(x_0)}{x - x_0}$$

$$= f(x_0)g'(x_0) + f'(x_0)g(x_0)$$

as required.

(iv) Let h be the function $1/g$ (with domain restricted to those x in g's domain with $g(x) \neq 0$). Then as in exercise 3 above h is differentiable at x_0 with

$$h'(x_0) = -\frac{g'(x_0)}{(g(x_0))^2}$$

Now we can apply the previous part to the functions f and h to give that hf ($=f/g$) is differentiable at x_0 with

$$\left(\frac{f}{g}\right)'(x_0) = \left(\frac{1}{g} \times f\right)'(x_0) = (hf)'(x_0)$$

$$= \underbrace{h(x_0)}_{\substack{\| \\ g(x_0)}}f'(x_0) + \underbrace{h'(x_0)}_{\substack{\| \\ -\frac{g'(x_0)}{(g(x_0))^2}}}f(x_0)$$

$$= \frac{g(x_0)f'(x_0) - g'(x_0)f(x_0)}{(g(x_0))^2}$$

as required.   □

No doubt those rules of differentiation are very familiar to you. Together they enable you to differentiate quite complicated functions; e.g.

$$f(x) = \frac{x \sin x}{x^2 + 1} \quad \Rightarrow$$

$$f'(x) = \frac{(x^2 + 1)(x \cos x + \sin x) - 2x.x \sin x}{(x^2 + 1)^2} \quad \text{etc.}$$

But the most striking way of making complicated functions by combining two others is by forming the composite 'function of a function'. For example you probably know that

$$h(x) = \underbrace{\sin(x^2)}_{f \ (g(x))} \quad \Rightarrow \quad h'(x) = \underbrace{2x}_{g'(x)} \underbrace{\cos(x^2)}_{f' \ (g(x))}$$

We can now establish this general result, commonly known as the 'chain rule':

Theorem   (*The chain rule*) Let g be differentiable at x_0 and let f be differentiable at $g(x_0)$. Then the composite function $f \circ g$ is differentiable at x_0 and its derivative is given by

$$(f \circ g)'(x_0) = g'(x_0) \times f'(g(x_0))$$

Proof   We are given that

g is differentiable at x_0 i.e. if

$$\underbrace{x_1, x_2, x_3, \ldots}_{\text{in } g\text{'s domain, } \neq x_0} \to x_0$$

then (by the theorem on page 148) g is also continuous at x_0 and so we have

$$g(x_1), g(x_2), g(x_3), \ldots \to g(x_0)$$

and (by the differentiability itself) as $n \to \infty$ we have

$$\frac{g(x_n) - g(x_0)}{x_n - x_0} \to g'(x_0)$$

f is differentiable at $g(x_0)$ i.e. if

$$\underbrace{y_1, y_2, y_3, \ldots}_{\text{in } f\text{'s domain, } \neq g(x_0)} \to g(x_0)$$

then as $n \to \infty$

$$\frac{f(y_n) - f(g(x_0))}{y_n - g(x_0)} \to f'(g(x_0))$$

In particular if

$$\underbrace{g(x_1), g(x_2), g(x_3), \ldots}_{\text{in } f\text{'s domain, } \neq g(x_0)} \to g(x_0)$$

then as $n \to \infty$ we have

$$\frac{f(g(x_n)) - f(g(x_0))}{g(x_n) - g(x_0)} \to f'(g(x_0))$$

To show that $f \circ g$ is differentiable at x_0 with derivative $g'(x_0)f'(g(x_0))$ we need to show that

if $\underbrace{x_1, x_2, x_3, \ldots \rightarrow x_0}_{\text{in } f \circ g\text{'s domain, } \neq x_0}$

then $\dfrac{f(g(x_n)) - f(g(x_0))}{x_n - x_0} \rightarrow g'(x_0)f'(g(x_0))$ as $n \rightarrow \infty$

It is tempting to derive the convergence of this last sequence by multiplying together the two convergent sequences 'boxed' above. But the one on the right depends upon the $g(x_n)$s being different from $g(x_0)$, otherwise the denominator of $g(x_n) - g(x_0)$ would be inadmissible.

So let us assume for the moment that $x_1, x_2, x_3, \ldots \rightarrow x_0$ and that only a finite number of the terms $g(x_1), g(x_2), g(x_3), \ldots$ actually equal $g(x_0)$. Then eventually (from some particular term onwards) all the $g(x_n)$s *will* be different from $g(x_0)$ and we *can* multiply together the two boxed sequences above to give

$$\frac{f(g(x_n)) - f(g(x_0))}{x_n - x_0} = \frac{g(x_n) - g(x_0)}{x_n - x_0} \times \frac{f(g(x_n)) - f(g(x_0))}{g(x_n) - g(x_0)}$$

$$\rightarrow g'(x_0) \times f'(g(x_0))$$

as required.

That argument works for the vast majority of functions but it's not quite foolproof because there exists the odd perverse function g and sequence

$\underbrace{x_1, x_2, x_3, \ldots \rightarrow x_0}_{\text{in } f \circ g\text{'s domain, } \neq x_0}$

with $g(x_{k_1}) = g(x_{k_2}) = g(x_{k_3}) = \cdots = g(x_0)$. In this case $g'(x_0)$ must be 0 because

$$\underset{\|}{\dfrac{g(x_{k_1}) - g(x_0)}{x_{k_1} - x_0}}, \quad \underset{\|}{\dfrac{g(x_{k_2}) - g(x_0)}{x_{k_2} - x_0}}, \quad \underset{\|}{\dfrac{g(x_{k_3}) - g(x_0)}{x_{k_3} - x_0}}, \ldots \rightarrow g'(x_0)$$

$$0 \qquad\qquad 0 \qquad\qquad 0$$

So to show that $(f \circ g)'(x_0) = g'(x_0) \times f'(g(x_0)) = 0$ we must simply show that

$$\frac{f(g(x_n)) - f(g(x_0))}{x_n - x_0} \rightarrow 0$$

But the terms of this sequence for which $g(x_n) \neq g(x_0)$ certainly do tend to $g'(x_0) \times f'(g(x_0))$ $(= 0)$ by the argument we used above, and the terms for which $g(x_n) = g(x_0)$ *are already* 0! So even in this perverse case $f \circ g$ is differentiable at x_0 with derivative $g'(x_0)f'(g(x_0))$. $\qquad\square$

Perhaps it is now time to pause to consider the alternative notation of 'dy/dx' sometimes called the 'Leibniz notation' (after Gottfried Leibniz who, along with Isaac Newton, is jointly credited with developing the differential calculus in the late 17th century). If we have a function f and write $y = f(x)$ then

$$\frac{f(x) = f(x_0)}{x - x_0} = \frac{\text{change in } y}{\text{change in } x} = \frac{\delta y}{\delta x}$$

where 'δ' is shorthand for 'change in'. Hence

$$f'(x_0) = \lim_{x \to x_0} \frac{f(x) - f(x_0)}{x - x_0} = \lim_{\delta x \to 0} \frac{\delta y}{\delta x}$$

and this leads to the notation of dy/dx for $f'(x_0)$ (a major shortcoming of the notation being that the x_0 does not feature in it!). It must be stressed that 'dy/dx' is a single unsplittable entity – the 'dy' and 'dx' are meaningless by themselves.

But the notation is quite neat and useful because it often behaves as if dy/dx *were* a fraction, as the following examples illustrate:

I. If $f(x) = y$ and f has an inverse function $x = g(y)$ then, as in exercise 3 above,

$$g'(y) = \frac{1}{f'(x)}$$

In the alternative notation this would be written

$$\frac{dx}{dy} = \frac{1}{dy/dx}$$

II. If $z = f(y)$ where $y = g(x)$ then $z = f(g(x))$ and by the above theorem

$$(f \circ g)'(x) = f'(g(x)) \times g'(x)$$

In the alternative notation this would be written

$$\frac{dz}{dx} = \frac{dz}{dy} \times \frac{dy}{dx}$$

So although dy/dx is *not* a fraction it does sometimes seem to behave like one.

The results of these last two sections form the 'bread and butter' results of differentiation, probably familiar to you from your school studies. But this is just the first step in our analytical development of differentiation which we shall pursue after the next exercises.

Exercises

1   Let $a > 0$ with $a \neq 1$.
   (i) Use the fact that $a^x = e^{x \log a}$ (exercise 2 page 118) to show that if $f(x) = a^x$ $(x \in \mathbb{R})$ then f is differentiable with derivative given by $f'(x) = a^x \log a$.
   (ii) Use the fact that

$$\log_a x = \frac{\log x}{\log a}$$

(exercise 2 page 118 again to show that if $g(x) = \log_a x$ $(x > 0)$ then g is differentiable with derivative given by $g'(x) = 1/(x \log a)$.

2   Let the functions f and g be given by

$$f(x) = \cosh^{-1}(x/2) \quad (x > 2) \quad \text{and} \quad g(x) = \log(x + \sqrt{(x^2 - 4)}) \quad (x > 2)$$

Use our various results concerning differentiation to show that f and g are differentiable with $f' = g'$. Does this imply that f and g are equal? What do you think it *does* imply about f and g? (We shall actually deduce the necessary theorems in the next section.)

3   (i) The function

$$y = F(x) = \sqrt{(1 - x^2)} \qquad x \in (-1, 1)$$

can be expressed in the alternative 'parametric' form

$$x = f(t) = \cos t \qquad y = g(t) = \sin t \qquad t \in (0, \pi)$$

Find $F'(x)$ and verify that it equals $g'(t)/f'(t)$.
   (ii) In general suppose that the function $y = F(x)$ is expressed in parametric form $x = f(t)$, $y = g(t)$ where f and g are differentiable and $f'(t) \neq 0$. Show that

$$F'(x) = \frac{g'(t)}{f'(t)}$$

Express this result in terms of dy/dx etc. and note that once again this *appears* to behave as a sensible fraction.

4   Let the functions f, g and h be given by

$$f(x) = \begin{cases} \sin(1/x) & x \neq 0 \\ 0 & x = 0 \end{cases}$$

$$h(x) = \begin{cases} x^2 \sin(1/x) & x \neq 0 \\ 0 & x = 0 \end{cases} \qquad g(x) = \begin{cases} x \sin(1/x) & x \neq 0 \\ 0 & x = 0 \end{cases}$$

(as in exercise 5 on page 132). Show that all three functions are differentiable for $x \neq 0$ but that only h is differentiable at $x = 0$.

5   When working out limits of the type $\lim_{x \to x_0}(f(x)/g(x))$ we can use the fact that limits behave in a sensible way to deduce that the

answer is $\lim_{x \to x_0} f(x)$ divided by $\lim_{x \to x_0} g(x)$. But if both these limits are 0 it is impossible to draw any conclusions about the limit of the quotient. However derivatives now help us to solve this problem.

Let f and g be functions which are both differentiable at x_0 with $f(x_0) = g(x_0) = 0$ and with $g'(x_0) \neq 0$. Assume that the domains of f and g are suitably overlapping so that $\lim_{x \to x_0} (f(x)/g(x))$ can be considered. Show that

$$\lim_{x \to x_0} \frac{f(x)}{g(x)} = \frac{f'(x_0)}{g'(x_0)}$$

Evaluate the following limits

(i) $\lim_{x \to 1} \dfrac{e^x}{x^2 + 1}$ (ii) $\lim_{x \to 0} \dfrac{\sin x}{e^x - 1}$

We extend this method further in the next exercises.

(Solutions on page 264)

A mean theorem

After learning how to differentiate at school you then probably used differentiation to help you sketch graphs and to find maxima and minima of functions.

Example Let the function f be given by

$$f(x) = (x - 1)^2(x - 7)   x \in [0, 8]$$

Differentiate f and sketch its graph.
Find the maximum and minimum of f.

Solution By using the rules of differentiation established in the previous theorem we see that

$$f'(x) = (x - 1)^2 + 2(x - 1)(x - 7) = 3(x - 1)(x - 5)$$

So the derivative is zero when $x = 1$ and when $x = 5$. The value of the function at these points is 0 and -32 respectively and so we can begin to sketch the graph by marking a zero gradient at the points $(1, 0)$ and $(5, -32)$. Noting the two extreme points $(0, -7)$ and $(8, 49)$ and the other point $(7, 0)$ where the graph crosses the x-axis gives the situation shown on the left in the next figure. Since f is a nicely-behaved (differentiable) function we are immediately able to fill in a fairly good sketch of the graph as shown on the right.

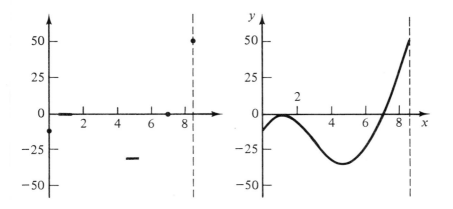

You may recall from the theorem on page 140 that any continuous function defined on a closed bounded interval has a maximum and a minimum. In this example it is clear that the maximum is 49 which occurs at $x = 8$ and the minimum is -32 which occurs at $x = 5$. □

That example begins to illustrate the role of points where a function f has zero gradient – called the *stationary points* (or *turning points*) of f. In the above example the function reaches a temporary hilltop at $x = 1$, in other words $f(x) \leqslant f(1)$ in some small interval around $x = 1$: f is said to have a 'local maximum' there. In general a function f has a *local maximum* at x_0 if $f(x) \leqslant f(x_0)$ for each x in some interval around x_0. Similarly f has a *local minimum* at x_0 if $f(x) \geqslant f(x_0)$ for each x in some interval around x_0: in the above example f has a local minimum at $x = 5$.

As you might suspect, if a differentiable function has a local maximum or minimum then its derivative is zero there; i.e. for differentiable functions local maxima and minima occur at stationary points, as we now prove. (In addition some stationary points are neither local maxima nor local minima, and we shall see some examples in the next exercises.)

> **Theorem**   Let f be a differentiable function. Then at any local maximum or minimum the derivative of f is 0; i.e. local maxima and minima occur at stationary points.

> **Proof**   If f has a local maximum at x_0 then by the definition $f(x) \leqslant f(x_0)$ for each x in some interval I around x_0. In other words there are

xs in I to the left of x_0 and   xs in I to the right of x_0

For all such x we have $x < x_0$ and $f(x) \leqslant f(x_0)$. Hence	For all such x we have $x > x_0$ and $f(x) \leqslant f(x_0)$. Hence
$$\frac{f(x) - f(x_0)}{x - x_0} \geqslant 0$$	$$\frac{f(x) - f(x_0)}{x - x_0} \leqslant 0$$
(This is reminiscent of exercise 4 on page 142 where we saw that for points Q on a graph to the left of the maximum at P the chord PQ has gradient $\geqslant 0$.)	(This is reminiscent of exercise 4 on page 142 where we saw that for points Q on a graph to the right of the maximum at P the chord PQ has gradient $\leqslant 0$.)
So **if**	So **if**
$$\underbrace{x_1, x_2, x_3, \ldots}_{< x_0} \to x_0$$	$$\underbrace{x_1, x_2, x_3, \ldots}_{> x_0} \to x_0$$
then	**then**
$$\underbrace{\frac{f(x_n) - f(x_0)}{x_n - x_0}}_{\geqslant 0} \to f'(x_0)$$	$$\underbrace{\frac{f(x_n) - f(x_0)}{x_n - x_0}}_{\leqslant 0} \to f'(x_0)$$
and it follows that $f'(x_0) \geqslant 0$.	and it follows that $f'(x_0) \leqslant 0$.

We can therefore deduce that $f'(x_0) = 0$ and that f has a stationary point at x_0 as claimed.

The verification for the local minimum is very similar (or it can be deduced by considering the local maximum of the function $-f$). $\square$

Of course, as the previous example shows, the local maximum of a function need not give its overall maximum – a local maximum is merely a temporary peak. But consider a continuous function f with domain $[a, b]$. By the theorem on page 140 this function has a maximum. If that maximum occurs at some $x_0 \in (a, b)$ then that clearly gives a local maximum of f. So the only places where such an f can achieve its maximum are at $x = a$ or at $x = b$ or at a local maximum. (We could of course make the corresponding deductions about the minimum.) Before using that simple observation to deduce the next theorem we illustrate the graphs of several such continuous functions and note where their maxima and minima occur:

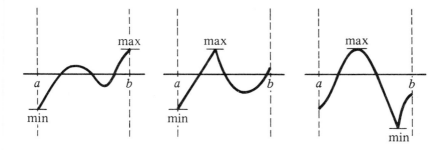

Theorem   Let $a < b$ and let f be a differentiable function with domain $[a, b]$ and with $f(a) = f(b)$. Then $f'(c) = 0$ for some $c \in (a, b)$.

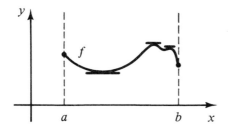

Proof   It is clear from the way in which we have approached this theorem that we must consider the maximum or minimum of f on $[a, b]$.

By the continuity of f on $[a, b]$ the function certainly has a maximum and a minimum (from the theorem on page 140). By our comments above the maximum is reached at $x = a$ or at $x = b$ or at a local maximum of f (with similar conclusions for the minimum). So there are three cases to consider:

Case   I: f's maximum is reached at a local maximum at
$c \in (a, b)$ ⎫ (these could
Case II: f's minimum is reached at a local minimum at ⎬ both occur)
$d \in (a, b)$ ⎭
Case III: cases I and II fail so that f's maximum and minimum are both
reached at $x = a$ and $x = b$.

In case I (see left-hand figure below) f has a local maximum at $c \in (a, b)$ and so **as f is differentiable at $c \in (a, b)$** we can use the previous theorem to deduce that $f'(c) = 0$ as required.

A very similar argument works for case II.

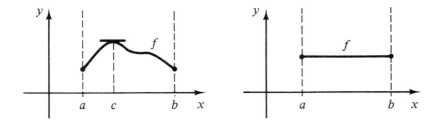

In case III (see right-hand figure above) f takes no value higher than $f(a)$ ($=f(b)$) and no value lower than $f(a)$. Hence in this case $f(x) = f(a)$ for each $x \in [a, b]$ and $f'(c) = 0$ for each c in this case.

Therefore in all cases we have found a $c \in (a, b)$ with $f'(c) = 0$, and the theorem is proved.   □

Although that was a theorem about a differentiable function f we didn't actually use the differentiability of f throughout $[a, b]$. In fact the only properties which we used (and which were highlighted by bold print in the proof) were that f was continuous on $[a, b]$ and differentiable at each point in (a, b). Although it's more of an effort to spell out these conditions it suits pure mathematicians who like to state theorems in as general a form as possible. Bearing in mind those comments we are now able immediately to deduce the following extension of the theorem; it is known as Rolle's theorem, first published by the French mathematician Michel Rolle in 1691.

> ***Corollary*** (*Rolle's theorem*) Let $a < b$ and let f be a continuous function with domain $[a, b]$, which is differentiable at each point in (a, b), and which has $f(a) = f(b)$. Then $f'(c) = 0$ for some $c \in (a, b)$.   □

That is a genuine (if slight) extension of the theorem because, for example, the corollary will apply to the function f given by

$$f(x) = \begin{cases} x\sin(1/x) & 0 < x \leqslant 1/\pi \\ 0 & x = 0 \end{cases}$$

(because, rather as we've seen in the exercises, this f is continuous on $[0, 1/\pi]$, differentiable on $(0, 1/\pi)$ and $f(0) = f(1/\pi)$). However, the theorem cannot be applied to f because (again as in the exercises) f is not differentiable at $x = 0$.

This section is called 'a mean theorem' and that's what we're coming to next: it's quite a natural result which will be referred to many times in advanced analysis courses. By way of an illustrative example imagine that you have just been on a drive and at the end of the journey you calculate the mean (or average) speed of the journey

$$\frac{\text{total distance travelled}}{\text{total time taken}}$$

Suppose that this turns out to be 40 mph. Then at some stage of the journey you must have been travelling **at** 40 mph. In this particular example that is not a surprising conclusion because to average 40 mph your speed at some stage must have exceeded 40 mph and so at some time your speedometer passed through the 40 mph mark. Consider now the distance/time graph of your journey: the gradient at any point is the speed at that time, and the mean (or average) speed is the gradient of the straight line shown.

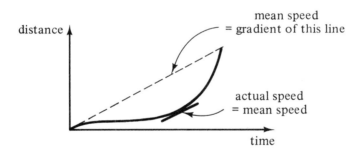

We are now able to state the general result, known as 'the mean value theorem'.

> ***Theorem***   (*The mean value theorem*) Let $a < b$ and let f be a continuous function with domain $[a, b]$ and which is differentiable at each point in (a, b). Then for some $c \in (a, b)$
>
> $$f'(c) = \frac{f(b) - f(a)}{b - a} \quad (= \text{mean or average change in } f \text{ from } a \text{ to } b)$$

> ***Proof***   Compare the conditions in this theorem with that in Rolle's theorem (the previous corollary). You will see that they are the same except that we have now dropped the requirement that $f(a) = f(b)$.

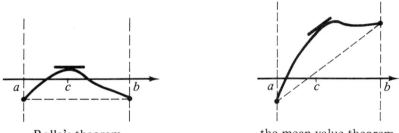

Rolle's theorem   the mean value theorem

In some sense the mean value theorem is merely a tilted version of Rolle's theorem and to get back from f to a more 'level' function g to which Rolle's theorem can be applied we define

$$g(x) = f(x) - (x - a)\frac{f(b) - f(a)}{b - a} \quad x \in [a, b]$$

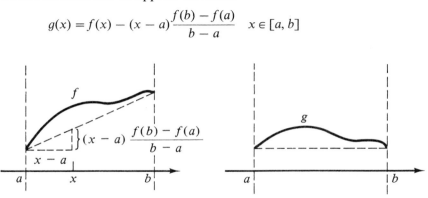

It is clear that g is a simple arithmetic combination of functions which are continuous on $[a, b]$ and differentiable on (a, b) and so g too has these properties. But in addition

$$g(a) = f(a) - (a - a)\frac{f(b) - f(a)}{b - a} = f(a)$$

and

$$g(b) = f(b) - (b - a)\frac{f(b) - f(a)}{b - a} = f(b) - (f(b) - f(a)) = f(a)$$

and so $g(a) = g(b)$ as hoped for.

Hence g satisfies all the requirements of Rolle's theorem and we can deduce that $g'(c) = 0$ for some $c \in (a, b)$. But by our standard techniques of differentiation

$$g'(x) = f'(x) - \frac{f(b) - f(a)}{b - a}$$

and in particular

$$g'(c) = f'(c) - \frac{f(b) - f(a)}{b - a} = 0 \quad \text{and so} \quad f'(c) = \frac{f(b) - f(a)}{b - a}$$

as required.   □

We know that if f is a constant function (i.e. $f(x) = k$ for each x, where k is some fixed number) then $f'(x) = 0$. We are now able to use the mean value theorem to deduce the converse, namely that if f's domain is an interval and $f'(x) = 0$ for each x then f is a constant function.

> ***Corollary*** Let f be a differentiable function whose domain D is an interval and such that $f'(x) = 0$ for each $x \in D$. Then f is a constant function; i.e. there is some number k with $f(x) = k$ for each $x \in D$.

Proof Let a and b be any two numbers in D, with $a < b$ say. We shall use the mean value theorem to show that $f(a) = f(b)$: since a and b were arbitrary it will then follow that $f(x)$ takes the same value for each $x \in D$.

So note that f is differentiable on $[a, b]$ (although we only actually need continuity on $[a, b]$ and differentiability on (a, b)). By the mean value theorem applied to f on $[a, b]$ we know that there exists $c \in (a, b)$ with

$$f'(c) = \frac{f(b) - f(a)}{b - a}$$

But $f'(x) = 0$ for each x and so in particular $f'(c) = 0$. Hence

$$\frac{f(b) - f(a)}{b - a} = 0 \quad \text{and} \quad f(a) = f(b)$$

as claimed. As we commented above, this shows that f always takes the same value and is a constant function.   □

Exercises

1   We have seen that if f is a function then so is f' (perhaps with a smaller domain). So we can consider the differentiability of the function f' and work out its derivative $(f')'$ of f'' (or d^2y/dx^2): this is called the *second derivative* of f. In a similar way we can repeatedly differentiate f and find its nth derivative denoted by $f^{(n)}$ (or $d^n y/dx^n$).

(i) For each positive integer n find the nth derivative of the functions

$$f(x) = x^4 \quad x \in \mathbb{R} \quad \text{and} \quad g(x) = \sin x \quad x \in \mathbb{R}$$

(ii) Let $f(x) = x^2(x^2 - 2)$ $(x \in \mathbb{R})$. Show that f has stationary points at $x = -1, 0$ and 1.

Sketch the graph of f.

Show also that at f's local maximum f'' is negative and at each of its local minima f'' is positive.

(iii) Let f and h be the functions we have met before which are defined by

$$f(x) = x^3 \quad x \in \mathbb{R} \qquad\qquad h(x) = \begin{cases} x^2 \sin(1/x) & x \neq 0 \\ 0 & x = 0 \end{cases}$$

Show that f and h each has a stationary point at $x = 0$ but that neither of these stationary points is a local maximum or minimum.

Show that the graph of f'' actually crosses the x-axis at 0 (f is said to have a *point of inflection* at $x = 0$).

(You may think that at a stationary point x_0 of a function f

$$f''(x_0) < 0 \Rightarrow f \text{ has a local maximum at } x_0$$
$$f''(x_0) > 0 \Rightarrow f \text{ has a local minimum at } x_0$$
$$f''(x_0) = 0 \Rightarrow f \text{ has a point of inflection at } x_0$$

But that's not entirely true and we shall learn the full story in the next section.)

2   Let f and g be functions with suitably-overlapping domains and assume that each of the functions can be differentiated n times. Use the principle of mathematical induction to show that the product fg is differentiable n times and that

$$(fg)^{(n)} = \binom{n}{0} fg^{(n)} + \binom{n}{1} f'g^{(n-1)} + \binom{n}{2} f''g^{(n-2)} + \cdots$$
$$+ \binom{n}{r} f^{(r)}g^{(n-r)} + \cdots + \binom{n}{n} f^{(n)}g$$

where $\binom{n}{r}$ is the 'binomial coefficient' $n!/(r!(n-r)!)$. This is known as *Leibniz' rule*.

Find $(fg)^{(7)}(x)$ where $f(x) = x^2$ and $g(x) = e^{2x}$ $(x \in \mathbb{R})$.

3   Apply the mean value theorem to the function f given by

$$f(x) = \sin^{-1} x \qquad x \in [-1, 1]$$

between 0 and x to deduce that

$$x \leqslant \sin^{-1} x \leqslant \frac{x}{\sqrt{(1 - x^2)}}$$

for each $x \in [0, 1)$.

4   Let f and g be functions which are continuous on $[a, b]$ and differentiable on (a, b), where $a < b$, and such that $g'(x)$ is never zero. We wish to prove that there exists $c \in (a, b)$ with

$$\frac{f'(c)}{g'(c)} = \frac{f(b) - f(a)}{g(b) - g(a)}$$

What's wrong with the following 'proof'?

'Applying the mean value theorem to f and to g shows that there exists $c \in (a, b)$ with

$$f'(c) = \frac{f(b) - f(a)}{b - a} \quad \text{and} \quad g'(c) = \frac{g(b) - g(a)}{b - a}$$

Dividing the left-hand expression by the right-hand one shows that

$$\frac{f'(c)}{g'(c)} = \frac{f(b) - f(a)}{g(b) - g(a)}$$

as required.'

By considering the function h given by

$$h(x) = f(x) - \frac{f(b) - f(a)}{g(b) - g(a)} g(x) \quad x \in [a, b]$$

find a valid proof of the above result.

5   We saw in exercise 5 on page 157 that if f and g are differentiable at x_0 (with suitably-overlapping domains) with $f(x_0) = g(x_0) = 0$ and $g'(x_0) \neq 0$ then

$$\lim_{x \to x_0} \frac{f(x)}{g(x)} = \frac{f'(x_0)}{g'(x_0)}$$

But what if $g'(x_0) = 0$? We can use the previous exercise to extend this result. So assume now that f and g are differentiable in an interval around x_0 with $f(x_0) = g(x_0) = 0$ and that

$$\lim_{x \to x_0} \frac{f'(x)}{g'(x)}$$

exists. Then show that

$$\lim_{x \to x_0} \frac{f(x)}{g(x)} = \lim_{x \to x_0} \frac{f'(x)}{g'(x)}$$

This is known as *L'Hopital's rule* (after the French mathematician Guillaume L'Hopital who published it in 1696, although he apparently learnt it from Johann Bernoulli). Use it to evaluate the following limits, where α is a fixed number.

(i) $\lim\limits_{x \to 1} \dfrac{\log x}{x - 1}$

(ii) $\lim\limits_{x \to 0} \dfrac{\cos x - 1}{x^2}$

(iii) $\lim\limits_{x \to 0} \dfrac{\log(1 + \alpha x)}{x}$

(iv) $\lim\limits_{n \to \infty} \left(1 + \dfrac{\alpha}{n}\right)^n$

6   As we defined earlier a function f is strictly increasing if whenever $a < b$ in f's domain it follows that $f(a) < f(b)$: similarly a function is *increasing* if $a < b$ implies that $f(a) \leqslant f(b)$ (with corresponding definitions of 'strictly decreasing' and 'decreasing').

Now let f be a differentiable function whose domain is an interval. Show that:

(i) if $f'(x) > 0$ for each x then f is strictly increasing;
(ii) if $f'(x) \geqslant 0$ for each x then f is increasing;
(iii) if $f'(x) < 0$ for each x then f is strictly decreasing;
(iv) if $f'(x) \leqslant 0$ for each x then f is decreasing.

Deduce that $\sin x \leqslant x$ for each $x \geqslant 0$.

7   (i) Let f and g be differentiable functions with domain an interval D and with $f' = g'$ (i.e. $f'(x) = g'(x)$ for each $x \in D$). Show that f and g differ by a constant; i.e. that there exists a number k with $f(x) = g(x) + k$ for each $x \in D$.

(ii) Use the result of exercise 2 on page 157 to show that

$$\cosh^{-1}(x/2) = \log(x + \sqrt{(x^2 - 4)}) - \log 2 \quad \text{for each } x \geqslant 2$$

8   Let f be a function with domain $(0, \infty)$ and range $\mathbb{R}$ and let g be a function with domain $\mathbb{R}$ and range $(0, \infty)$. Assume that

$$f'(x) = 1/x \quad x > 0 \quad \text{and} \quad g'(x) = g(x) \quad x \in \mathbb{R}$$

Use exercise 6 to show that g is strictly increasing (and hence that it has an inverse). Then use exercise 4 on page 150 to find the derivative of g's inverse, and show that g's inverse and f differ by a constant. Hence

(i) give an alternative proof that the log and exp functions are inverses of each other;

(ii) Show that $g(x) = e^{x+k}$ for some constant k.

(Solutions on page 267)

Polynomial approximations

For some applications where only approximate answers are required it is possible to approximate to a function using a much more elementary expression. For example in the approximation

$$\sin x \approx x - \tfrac{1}{6}x^3$$

the left-hand and right-hand sides differ by less than 0.02 for $x \in [-\pi/3, \pi/3]$:

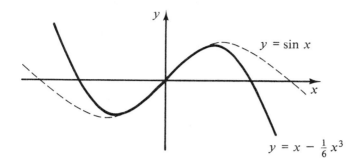

If a closer approximation to the sine function is required over a larger domain then a better approximation is

$$\sin x \approx x - \tfrac{1}{6}x^3 + \tfrac{1}{120}x^5$$

where here the two sides differ by less than 0.005 for any $x \in [-\pi/2, \pi/2]$.

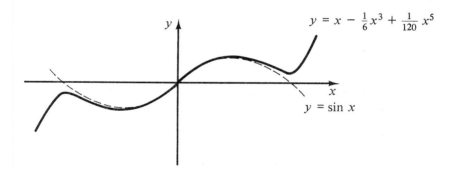

But how are the expressions $x - \tfrac{1}{6}x^3$ and $x - \tfrac{1}{6}x^3 + \tfrac{1}{120}x^5$ chosen?

A *polynomial* is a function of the form

$$f(x) = a_0 + a_1 x + a_2 x^2 + \cdots + a_n x^n \quad x \in \mathbb{R}$$

where the numbers $a_0, a_1, \ldots, a_n$ are constants: such a function is clearly differentiable arbitrarily often. If $a_n \neq 0$ then the polynomial is of the *n*th *degree*. Suppose that you knew that a particular function f was a polynomial and that you could calculate the value of f and any of its derivatives at any point. How would you calculate the coefficients $a_0, a_1, \ldots, a_n$?

Example Let f be a polynomial for which
$$f(0) = 1 \qquad f'(0) = -1 \qquad f''(0) = 4 \qquad f^{(3)}(0) = 12$$
$$\text{and} \quad f^{(n)}(0) = 0 \quad \text{for} \quad n \geqslant 4$$

Write f as a polynomial, calculating each of its coefficients specifically.

Solution Note firstly that if f is the polynomial
$$f(x) = a_0 \quad + a_1 x \qquad + a_2 x^2 \qquad + \cdots + a_n x^n$$

then
$$f'(x) = a_1 \quad + 2a_2 x \qquad + 3a_3 x^2 \qquad + \cdots + na_n x^{n-1}$$
$$f''(x) = 2!a_2 + 3 \times 2a_3 x \quad +4 \times 3a_4 x^2 \qquad + \cdots + n(n-1)a_n x^{n-2}$$
$$f^{(3)}(x) = 3!a_3 + 4 \times 3 \times 2a_4 x + 5 \times 4 \times 3a_5 x^2 + \cdots + n(n-1)(n-2)\, a_n x^{n-3}$$
$$\vdots$$

and so putting $x = 0$ we see that
$$f(0) = a_0, \quad f'(0) = a_1, \quad f''(0) = 2!a_2, \quad f^{(3)}(0) = 3!a_3, \quad f^{(4)}(0) = 4!a_4, \ldots$$

Hence
$$a_0 = f(0), \quad a_1 = f'(0), \quad a_2 = \frac{f''(0)}{2!}, \quad a_3 = \frac{f^{(3)}(0)}{3!}, \quad a_4 = \frac{f^4(0)}{4!}, \ldots$$

and in this case $a_0 = 1$, $a_1 = -1$, $a_2 = 4/2 = 2$, $a_3 = 12/6 = 2$ and all the other coefficients are zero. Hence f is the polynomial
$$f(x) = 1 - x + 2x^2 + 2x^3 \qquad \qquad \square$$

Now if we have a complicated function, then for many applications we would rather that f was expressed as a polynomial. Let us assume that the domain of the function f is an interval including 0 and that the function can be differentiated as often as we like. We might proceed by *assuming* that f is in fact a polynomial of degree 3 (i.e. a cubic) and, as in the above example, calculate the coefficients to give
$$f(x) \approx a_0 + a_1 x + a_2 x^2 + a_3 x^3$$
$$\uparrow$$
hopefully!

where $a_0 = f(0)$, $a_1 = f'(0)$, $a_2 = f''(0)/2!$ and $a_3 = f^{(3)}(0)/3!$.

Example Carry out the above procedure to find a polynomial of degree 3 which hopefully approximates to the sine function.

Repeat the procedure to find some polynomials of higher degrees which hopefully approximate even better to the sine function.

Solution If $f(x) = \sin x$ then
$$f'(x) = \cos x \quad f''(x) = -\sin x \quad f^{(3)}(x) = -\cos x \quad f^{(4)}(x) = \sin x$$
$$f^{(5)}(x) = \cos x \quad f^{(6)}(x) = -\sin x \quad f^{(7)}(x) = -\cos x \quad f^{(8)}(x) = \sin x \text{ etc.}$$

and so the above procedure yields the cubic polynomial with coefficients

$$a_0 = f(0) = 0 \quad a_1 = f'(0) = 1 \quad a_2 = \frac{f''(0)}{2!} = 0 \quad a_3 = \frac{f^{(3)}(0)}{3!} = \frac{-1}{6}$$

i.e.

$$x - \tfrac{1}{6}x^3$$

We have already seen that this cubic does indeed approximate to $\sin x$ for small x.

Continuing the procedure to find a polynomial of degree 5 which hopefully approximates to the sine function yields

$$a_0 = 0 \quad a_1 = 1 \quad a_2 = 0 \quad a_3 = -\tfrac{1}{6} \quad a_4 = \frac{f^{(4)}(0)}{4!} = 0$$

$$a_5 = \frac{f^{(5)}(0)}{5!} = \frac{1}{120}$$

i.e.

$$x - \tfrac{1}{6}x^3 + \tfrac{1}{120}x^5$$

which we already know to be a better approximation to $\sin x$.

Continuing in this way we get a sequence of polynomials which hopefully will provide better and better approximations for the sine function:

$$x$$
$$x - \tfrac{1}{6}x^3$$
$$x - \tfrac{1}{6}x^3 + \tfrac{1}{120}x^5$$
$$x - \tfrac{1}{6}x^3 + \tfrac{1}{120}x^5 - \tfrac{1}{5040}x^7$$
$$x - \tfrac{1}{6}x^3 + \tfrac{1}{120}x^5 - \tfrac{1}{5040}x^7 + \tfrac{1}{362880}x^9$$
$$x - \tfrac{1}{6}x^3 + \tfrac{1}{120}x^6 - \tfrac{1}{5040}x^7 + \tfrac{1}{362880}x^8 - \tfrac{1}{39916800}x^{11}$$
$$\vdots$$

In fact the domain of $[-\pi, \pi]$ is sufficient for practical applications of the sine function, and throughout that domain the 11th degree polynomial above never differs from $\sin x$ by more than 0.0005.   □

So given any function f whose domain is an interval including 0 and which is differentiable arbitrarily often we can construct a sequence of polynomials

$$a_0 \qquad\qquad \text{where } a_0 = f(0)$$
$$a_0 + a_1 x \qquad\qquad \text{where } a_1 = f'(0)$$
$$a_0 + a_1 x + a_2 x^2 \qquad\qquad \text{where } a_2 = f''(0)/2!$$
$$a_0 + a_1 x + a_2 x^2 + a_3 x^3 \qquad\qquad \text{where } a_3 = f^{(3)}(0)/3!$$
$$\vdots \qquad\qquad \vdots$$
$$a_0 + a_1 x + a_2 x^2 + \cdots + a_n x^n \qquad\qquad \text{where } a_n = f^{(n)}(0)/n!$$
$$\vdots \qquad\qquad \vdots$$

But we have no way of telling in general whether these polynomials will provide good approximation to f; sometimes they will and sometimes they will not:

Example   Let f and g be the functions

$$f(x) = e^x \quad x \in \mathbb{R} \qquad\qquad g(x) = \begin{cases} e^{-1/x^2} & x \neq 0 \\ 0 & x = 0 \end{cases}$$

For each of these functions calculate the sequence of polynomials described above and observe that in one case the polynomials seem to give better and better approximations to the function but that in the other case they do not.

Solution   In the case of f the coefficients of the polynomials are easy to calculate but in the case of g they are much harder: we'll deal with f first.

$$\qquad\qquad\qquad\qquad\qquad\qquad\qquad\qquad\qquad\qquad \textit{polynomials}$$

$f(x) = e^x$	$f(0) = 1$	$a_0 = 1$	1
$f'(x) = e^x$	$f'(0) = 1$	$a_1 = 1$	$1 + x$
$f''(x) = e^x$	$f''(0) = 1$	$a_2 = \dfrac{1}{2!}$	$1 + x + \dfrac{1}{2!}x^2$
$f^{(3)}(x) = e^x$	$f^{(3)}(0) = 1$	$a_3 = \dfrac{1}{3!}$	$1 + x + \dfrac{1}{2!}x^2 + \dfrac{1}{3!}x^3$
$\vdots$	$\vdots$	$\vdots$	$\vdots$

As we saw in chapter 3 the exponential function can be written as an infinite sum or series

$$f(x) = e^x = \exp(x) = 1 + x + \frac{1}{2!}x^2 + \frac{1}{3!}x^3 + \cdots$$

and for any particular x the polynomials we've constructed above are just the partial sums of that series. So in some sense those polynomials certainly do provide better and better approximations for f.

Now let us consider g. For $x \neq 0$ g is a composition of differentiable functions and it can be differentiated using the chain rule. But at $x = 0$ we have to resort to the definition of the derivative to see that

$$g'(0) = \lim_{x \to 0} \frac{g(x) - g(0)}{x - 0} = \lim_{x \to 0} \frac{e^{-1/x^2}}{x}$$

But if

$$\underbrace{x_1, x_2, x_3, \ldots}_{\neq 0} \to 0$$

then

$$-\frac{1}{x_n^2} - \log|x_n| \to -\infty \quad \text{and} \quad \left| \frac{e^{-1/x_n^2}}{x_n} \right| = e^{-1/x_n^2 - \log|x_n|} \to 0$$

Hence the above expression for $g'(0)$ is 0 and g' is the function given by

$$g'(x) = \begin{cases} \dfrac{2}{x^3} e^{-1/x^2} & x \neq 0 \\ 0 & x = 0 \end{cases}$$

We could (with much effort) continue in this way to find g'', $g^{(3)}$ etc. and to show that each of g's derivatives is 0 at $x = 0$. Hence

		polynomials
$g(0) = 0$	$a_0 = 0$	0
$g'(0) = 0$	$a_1 = 0$	$0 + 0x = 0$
$g''(0) = 0$	$a_2 = 0$	$0 + 0x + 0x^2 = 0$
$g^{(3)}(0) = 0$	$a_3 = 0$	$0 + 0x + 0x^2 + 0x^3 = 0$
$\vdots$	$\vdots$	$\vdots$

These polynomials are all the same and certainly (apart from at $x = 0$) they do not provide improving approximations for g.   □

In the early eighteenth century several British mathematicians (including Colin Maclaurin, James Stirling and Brook Taylor) worked on these sorts of polynomials. The following theorem gives us some measure of the accuracy of the polynomial approximation to the function: it is a particular form of what is now known as *Taylor's theorem*, although it wasn't until the end of the century that mathematicians like Augustin-Louis Cauchy actually considered whether taking these polynomials of higher degree would eventually approximate arbitrarily closely to the original function. Indeed the form of 'Taylor's theorem' given here was in fact due to Joseph Lagrange in 1797.

> **Theorem**   (*Taylor's theorem* at 0) Let f be a function whose domain is an interval which includes 0 and let f be differentiable more than n times. Then let
>
> $$a_0 = f(0), a_1 = f'(0), a_2 = \frac{f''(0)}{2!}, a_3 = \frac{f^{(3)}(0)}{3!}, \dots, a_n = \frac{f^{(n)}(0)}{n!}$$
>
> It follows that for any x_0 in f's domain $f(x_0)$ and the polynomial
>
> $$a_0 + a_1 x_0 + a_2 x_0^2 + \cdots + a_n x_0^n$$
>
> differ by an amount equal to $f^{(n+1)}(c)x_0^{n+1}/(n+1)!$ for some c between 0 and x_0.

Proof   Let d be the difference between the actual value of the function $f(x_0)$ and the polynomial $a_0 + a_1 x_0 + a_2 x_0^2 + \cdots + a_n x_0^n$, and let g be the function (with the same domain as f) given by

$$g(x) = \left(f(x) + (x_0 - x)f'(x) + \frac{(x_0 - x)^2}{2!}f''(x) + \cdots + \frac{(x_0 - x)^n}{n!}f^{(n)}(x) \right)$$
$$+ \left(\frac{x_0 - x}{x_0} \right)^{n+1} d$$

Then g is differentiable on $[0, x_0]$ and it has been cunningly chosen to have three rather neat properties. Firstly, when its derivative is calculated practically all the terms cancel out giving

$$g'(x) = \quad \cancel{f'(x)} +$$
$$[(x_0 - x)f''(x) - \cancel{f'(x)}] +$$
$$\left[\frac{(x_0 - x)^2}{2!}f^{(3)}(x) - \cancel{\frac{2(x_0 - x)}{2!}}f''(x) \right] +$$
$$\vdots$$
$$\left[\frac{(x_0 - x)^n}{n!}f^{(n+1)}(x) - \cancel{\frac{n(x_0 - x)^{n-1}}{n!}}f^{(n)}(x) \right] -$$
$$\frac{(n+1)(x_0 - x)^n}{x_0^{n+1}} d$$
$$= \frac{(n+1)(x_0 - x)^n}{x_0^{n+1}} \left[\frac{x_0^{n+1}}{(n+1)!}f^{(n+1)}(x) - d \right]$$

Also

$$g(0) = \left(f(0) + x_0 f'(0) + \frac{x_0^2}{2!}f''(0) + \cdots + \frac{x_0^n}{n!}f^{(n)}(0) \right) + d$$
$$= (a_0 + a_1 x_0 + a_2 x_0^2 + \cdots + a_n x_0^n) + d = f(x_0)$$

and

$$g(x_0) = (f(x_0) + 0 + 0 + \cdots + 0) \quad + \quad 0 \times d = f(x_0)$$

Hence $g(0) = g(x_0)$ and we can apply Rolle's theorem to g to deduce that $g'(c) = 0$ for some $c \in (0, x_0)$; i.e.

$$\underbrace{\frac{(n+1)(x_0 - c)^n}{x_0^{n+1}}}_{\neq 0} \left[\frac{x_0^{n+1}}{(n+1)!} f^{(n+1)}(c) - d \right] = 0 \quad \text{and}$$

$$d = \frac{x_0^{n+1}}{(n+1)!} f^{(n+1)}(c)$$

as required.   □

The theorem does not of course say that the polynomial is necessarily a good approximation to the function but it does give us a way of calculating the difference between the two.

Example   Show that the polynomial

$$x - \frac{x^3}{3!} + \frac{x^5}{5!} - \cdots \pm \frac{x^n}{n!} \quad \text{(where } n \text{ is odd)}$$

differs from $\sin x$ by at most $\pi^{n+1}/(n+1)!$ throughout the interval $[-\pi, \pi]$.

Use your calculator to find a polynomial which differs from $\sin x$ by less than $0.000\,005$ throughout $[-\pi, \pi]$.

Solution   Let $f(x) = \sin x$. Then by the theorem the size (or modulus) of the difference between $f(x_0)$ and the given polynomial at x_0 is equal to

$$\left| \frac{x_0^{n+1}}{(n+1)!} f^{(n+1)}(c) \right|$$

for some c between 0 and x_0. Now since n is odd $f^{(n+1)}(c)$ is $\pm \sin c$ and this can only range between -1 and 1. Hence for $x_0 \in [-\pi, \pi]$ the biggest value that difference can take is $\pi^{n+1}/(n+1)!$ as claimed.

Now we saw in exercise 4 on page 63 that for any number α the sequence $(\alpha^n/n!)$ converges to 0. Therefore $\pi^{n+1}/(n+1)!$ can be made arbitrarily small by choosing n large enough. In particular my calculator shows that

$$\frac{\pi^{12}}{12!} \approx 0.001\,93 \qquad \frac{\pi^{14}}{14!} \approx 0.000\,105 \qquad \frac{\pi^{16}}{16!} \approx 0.000\,004\,3$$

Hence the polynomial

$$x - \frac{x^3}{3!} + \frac{x^5}{5!} - \frac{x^7}{7!} + \frac{x^9}{9!} - \frac{x^{11}}{11!} + \frac{x^{13}}{13!} - \frac{x^{15}}{15!}$$

differs from $\sin x$ by less than 0.000 005 throughout $[-\pi, \pi]$. That is an incredibly close approximation: a drawn graph of the polynomial would be indistinguishable from that of the sine function on the interval $[-\pi, \pi]$.

□

It seems from that example that the polynomials

$$x$$

$$x - \frac{x^3}{3!}$$

$$x - \frac{x^3}{3!} + \frac{x^5}{5!}$$

$$x - \frac{x^3}{3!} + \frac{x^5}{5!} - \frac{x^7}{7!}$$

$$\vdots$$

eventually give arbitrarily close approximation to $\sin x$. Rather than write out that repetitive list we introduce an abbreviated version and call the 'infinite polynomial'

$$x - \frac{x^3}{3!} + \frac{x^5}{5!} - \frac{x^7}{7!} + \frac{x^9}{9!} - \cdots + (-1)^m \frac{x^{2m+1}}{(2m+1)!} + \cdots$$

the 'Taylor series' ('about $x = 0$' – but don't worry about that yet) of the sine function.

In general, given a function f which is differentiable arbitrarily often and whose domain includes 0, the *Taylor series of f (about $x = 0$)* is the expression

$$f(0) + f'(0)x + \frac{f''(0)}{2!}x^2 + \frac{f^{(3)}(0)}{3!}x^3 + \cdots + \frac{f^{(n)}(0)}{n!}x^n + \cdots$$

At the moment that is nothing more than shorthand for the fact that any of the polynomials obtained by stopping that expression at some point might make a good approximation to $f(x)$ (with the above theorem giving some measure of how good that approximation is). Of course you will have realised that we are back in the world of series or infinite sums and we shall soon see the connections in the next section.

Before revising series we have one piece of unfinished business here.

We remarked earlier that a function f has a 'stationary' point when $f'(x) = 0$. As you probably know if $f''(x)$ is then positive the stationary point is a local minimum of the function and if $f''(x)$ is negative the stationary point is a local maximum. But what if $f''(x) = 0$? The following theorem tells a fuller story:

> **Theorem**   Let f be a function which is differentiable arbitrarily often in some interval surrounding the point $x = a$. Assume that for some $m > 1$
>
> $$f'(a) = f''(a) = f^{(3)}(a) = \cdots = f^{(m-1)}(a) = 0 \quad \text{and} \quad f^{(m)}(a) \neq 0$$
>
> (i.e. the mth derivative is the first non-zero one at a). Then
> (i) if m is even and $f^{(m)}(a) > 0$ then f has a local minimum at $x = a$;
> (ii) if m is even and $f^{(m)}(a) < 0$ then f has a local maximum at $x = a$;
> (iii) if m is odd then f has neither a local maximum nor a local minimum at $x = a$.

Proof   We shall for the moment restrict the proof to the case when $a = 0$: the proof for general a will follow instantly in the next exercises.

We know that $f^{(m)}(0) \neq 0$ and that $f^{(m)}$ is differentiable (and hence continuous) in an interval surrounding $x = 0$. Therefore if $f^{(m)}(0) > 0$ it follows that $f^{(m)}(x) > 0$ for each x in some interval I surrounding 0, and that if $f^{(m)}(0) < 0$ then $f^{(m)}(x) < 0$ for each x in some interval I surrounding 0 (exercise 5, page 107).

Now by the above version of Taylor's theorem in the case $n = m - 1$ we have for any $x \in I$

$$f(x) = f(0) + f'(0)x + \frac{f''(0)}{2!} x^2 + \cdots + \frac{f^{(m-1)}(0)}{(m-1)!} x^{m-1} + \frac{f^{(m)}(c)}{m!} x^m$$

$$= f(0) + \frac{f^{(m)}(c)}{m!} x^m$$

where c is some number between 0 and x. Therefore for any $x \in I$

$$f(x) - f(0) = \overbrace{\frac{f^{(m)}(c)}{m!}}^{\text{same sign as } f^{(m)}(0)} x^m$$

if m is even this is >0
for each $x \neq 0$ in I;
if m is odd this is <0 to
the left of $x = 0$ and >0 to the right of $x = 0$

It is now clear that:

(i) if m is even and $f^{(m)}(0) > 0$ then $f(x) - f(0) > 0$ (and $f(x) > f(0)$) for each $x \neq 0$ in I; i.e. f has a local minimum at $x = 0$;

(ii) if m is even and $f^{(m)} < 0$ then $f(x) - f(0) < 0$ (and $f(x) < f(0)$) for each $x \neq 0$ in I; i.e. f has a local maximum at $x = 0$;

(iii) if m is odd $f(x) > f(0)$ on one side of $x = 0$ and $f(x) < f(0)$ on the other side of $x = 0$; i.e. f has neither a local maximum nor a local minimum there.

That completes the proof when $a = 0$: the general case follows easily in one of the next exercises.   □

Example   Determine the nature of the stationary point at $x = 0$ of each of the following functions (with domain $\mathbb{R}$).

(i) $f(x) = x^4 - x^2$;   (ii) $g(x) = x^4 - x^6$;   (iii) $h(x) = x^6 - x^5$.

Solution   The derivatives are easy to calculate and you can soon find the first non-zero derivative at $x = 0$ in each case:

$f'(0) = 0$ and $f''(0) = -2$ (negatative) (even)	$g'(0) = g''(0)$ $= g^{(3)}(0) = 0$ and $g^{(4)}(0) = 24$ (positive) (even)	$h'(0) = h''(0)$ $= h^{(3)}(0)$ $= h^{(4)}(0) = 0$ and $h^{(5)}(0) = -120$ (odd)
$\therefore f$ has a local maximum at $x = 0$	$\therefore g$ has a local minimum at $x = 0$	$\therefore h$ has neither a local maximum nor minimum at $x = 0$

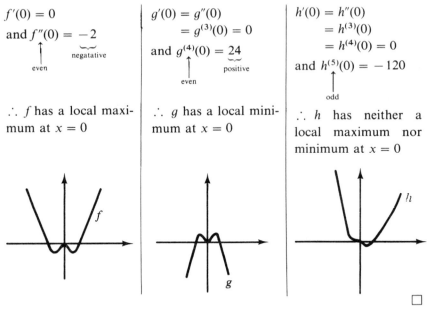

□

Exercises

1   Find the Taylor series (about $x = 0$) of the function cos and of the function cosh.

In each case suppose that we wish to find a polynomial which approximates to the function to within ε (>0) throughout the interval $[-k, k]$. Show that such a polynomial exists.

2   Let f be the function given by $f(x) = \log(1 + x)$ $(x > -1)$. Find the Taylor series of f (about $x = 0$).

Suppose that we wish to find a polynomial which approximates to f to within ε (>0) throughout the interval $[-\frac{1}{2}, 1]$. Show that such a polynomial exists.

3   Let g be a function which can be differentiated arbitrarily often and whose domain includes a. By applying the above form of Taylor's theorem to the function f given by $f(x) = g(x + a)$ show that the polynomial

$$g(a) + g'(a)(x_0 - a) + \frac{g''(a)}{2!}(x_0 - a)^2 + \cdots + \frac{g^{(n)}(a)}{n!}(x_0 - a)^n$$

differs from $g(x_0)$ by an amount equal to $g^{(n+1)}(c)(x_0 - a)^{n+1}/(n + 1)!$ for some c between a and x_0. (This is the more general form of Taylor's theorem.)

The expression

$$g(a) + g'(a)(x - a) + \frac{g''(a)}{2!}(x - a)^2 + \cdots + \frac{g^{(n)}(a)}{n!}(x - a)^n + \cdots$$

is called the *Taylor's series of g* (about $x = a$).

Find the Taylor series of the log function about $x = 1$.

4   Complete the proof of the last theorem (i.e. for general a).

Find the stationary points of the function given by $f(x) = x^3 e^x$ $(x \in \mathbb{R})$ and determine their nature. Hence sketch the graph of f.

(Solutions on page 272)

Series revisited

We originally defined the exponential function by the rule

$$\exp(x) = 1 + x + \frac{x^2}{2!} + \frac{x^3}{3!} + \cdots \qquad x \in \mathbb{R}$$

having noted that for any number x that series converged (i.e. added up to a sensible sum).

In general given any coefficients $a_0, a_1, a_2, \ldots$ we can define a *power series* by

$$a_0 + a_1 + a_2 x^2 + a_3 x^3 + \cdots$$

For any value of x this is a series (like we studied in chapter 2) which may or may not converge. It certainly converges when $x = 0$! So let D be the set of those x for which this series converges and define a function f by its sum; i.e.

$$f(x) = a_0 + a_1 x + a_2 x^2 + a_3 x^3 + \cdots \qquad x \in D$$

Examples of power series which we have seen before are

(1) $\qquad 1 + x + \dfrac{x^2}{2!} + \dfrac{x^3}{3!} + \dfrac{x^4}{4!} + \cdots$

which, as we've just commented, converges to e^x for each $x \in \mathbb{R}$.

(2) $\qquad 1 + x + x^2 + x^3 + \cdots$

which, as we saw in exercise 7 on page 74, converges to $(1 + x)^{-1}$ for $x \in (-1, 1)$.

In addition we can now see that a Taylor series is a particular example of a power series. For instance, as we began to see in the previous section, it turns out that the power series given by

(3) $\qquad x - \dfrac{x^3}{3!} + \dfrac{x^5}{5!} - \dfrac{x^7}{7!} + \cdots$

converges to $\sin x$ for each x.

So instead of just regarding it as a shorthand for a set of polynomials we can now see that a Taylor series of a function f is a power series. And we can use Taylor's theorem from the previous section to find those x for which the series actually converges to $f(x)$.

Example Find the Taylor series for the function f given by
$$f(x) = e^{-2x} \qquad x \in \mathbb{R}$$
and show that it converges to $f(x)$ for each $x \in \mathbb{R}$.

Solution Clearly
$$f'(x) = -2e^{-2x}, \quad f''(x) = 4e^{-2x}, \quad f^{(3)}(x) = -8e^{-2x}, \quad f^{(4)}(x) = 16e^{-2x}, \ldots$$
giving
$$f(0) = 1, \quad f'(0) = -2, \quad f''(0) = 4, \quad f^{(3)}(0) = -8, \quad f^{(4)}(0) = 16, \ldots$$
Hence the Taylor series for the function f is
$$1 - 2x + 2x^2 - \tfrac{4}{3}x^3 + \tfrac{2}{3}x^4 + \cdots + (-1)^{n-1}\dfrac{2^n}{n!}x^n + \cdots$$
Now we hope that for any number x_0 the series
$$1 - 2x_0 + 2x_0^2 - \tfrac{4}{3}x_0^3 + \tfrac{2}{3}x_0^4 + \cdots + (-1)^{n-1}\dfrac{2^n}{n!}x_0^n + \cdots$$

converges to $f(x_0)$. To test the convergence of a series we look at its partial sums

$$1$$
$$1 - 2x_0$$
$$1 - 2x_0 + 2x_0{}^2$$
$$1 - 2x_0 + 2x_0{}^2 - \tfrac{4}{3}x_0{}^3$$
$$1 - 2x_0 + 2x_0{}^2 - \tfrac{4}{3}x_0{}^3 + \tfrac{2}{3}x_0{}^4$$
$$\vdots$$

To show that the sequence converges to $f(x_0)$ we shall show that the sequence of differences

$$d_0 = f(x_0) - 1$$
$$d_1 = f(x_0) - (1 - 2x_0)$$
$$d_2 = f(x_0) - (1 - 2x_0 + 2x_0{}^2)$$
$$d_3 = f(x_0) - (1 - 2x_0 + 2x_0{}^2 - \tfrac{4}{3}x_0{}^3)$$
$$d_4 = f(x_0) - (1 - 2x_0 + 2x_0{}^2 - \tfrac{4}{3}x_0{}^3 + \tfrac{2}{3}x_0{}^4)$$
$$\vdots$$

converges to 0.

But by Taylor's theorem on page 174

$$d_n = \frac{x_0{}^{n+1}}{(n+1)!} f^{(n+1)}(c_n) = \frac{x_0{}^{n+1}}{(n+1)!} \times \pm 2^{n+1} e^{-2c_n}$$

for some c_n between 0 and x_0 (which may be positive or negative). Now each c_n must lie between $-|x_0|$ and $|x_0|$ and so e^{-2c_n} will be at most $e^{2|x_0|}$. Also, as we remarked above, the sequence $(\alpha^n/n!)$ converges to 0 for any number α. Hence as $n \to \infty$ we have

$$-e^{2|x_0|} \times \frac{|2x_0|^{n+1}}{(n+1)!} \quad \leqslant \quad d_n \quad \leqslant \quad e^{2|x_0|} \times \frac{|2x_0|^{n+1}}{(n+1)!}$$
$$\downarrow \qquad\qquad\qquad \downarrow \qquad\qquad\qquad\qquad \downarrow$$
$$0 \qquad\qquad \therefore \quad 0 \qquad\qquad\qquad\qquad 0$$

and the sequence (d_n) converges to 0 as required.

We have therefore shown that in this case the Taylor series for f actually converges to $f(x_0)$ for any x_0.   □

In that way we can find the Taylor series for each of the standard functions f and find the set of x for which the series converges to $f(x)$. The following table gives a few such series for reference and some more will be found in the exercises. By way of warning we also include the odd example from page 172: in that last example in the table the series converges for all $x \in \mathbb{R}$ but it only converges to $f(x)$ for $x = 0$.

Function f	Taylor series	The set of x for which the series converges to $f(x)$
$f(x) = e^x$	$1 + x + \dfrac{x^2}{2!} + \dfrac{x^3}{3!} + \dfrac{x^4}{4!} + \ldots$	$\mathbb{R}$
$f(x) = \sin x$	$x - \dfrac{x^3}{3!} + \dfrac{x^5}{5!} - \dfrac{x^7}{7!} + \cdots$	$\mathbb{R}$
$f(x) = \cos x$	$1 - \dfrac{x^2}{2!} + \dfrac{x^4}{4!} - \dfrac{x^6}{6!} + \cdots$	$\mathbb{R}$
$f(x) = \sinh x$	$x + \dfrac{x^3}{3!} + \dfrac{x^5}{5!} + \dfrac{x^7}{7!} + \cdots$	$\mathbb{R}$
$f(x) = \cosh x$	$1 + \dfrac{x^2}{2!} + \dfrac{x^4}{4!} + \dfrac{x^6}{6!} + \cdots$	$\mathbb{R}$
$f(x) = \log(1 + x)$	$x - \dfrac{x^2}{2} + \dfrac{x^3}{3} - \dfrac{x^4}{4} + \cdots$	$(-1, 1]$
$f(x) = (1 + x)^\alpha$	$1 + \alpha x + \dfrac{\alpha(\alpha - 1)}{2!} x^2$ $+ \dfrac{\alpha(\alpha - 1)(\alpha - 2)}{3!} x^3 + \cdots$	$\begin{cases} \mathbb{R} \text{ if } \alpha \geqslant 0 \text{ is an integer} \\ [-1, 1] \text{ for other } \alpha > 0 \\ (-1, 1] \text{ if } -1 < \alpha < 0 \\ (-1, 1) \text{ if } \alpha \leqslant -1 \end{cases}$
$f(x) = \begin{cases} e^{-1/x^2} & x \neq 0 \\ 0 & x = 0 \end{cases}$	$0 + 0x + 0x^2 + 0x^3 + \cdots$	$\{0\}$ only

Several properties of Taylor series begin to emerge when you construct a list of them. The first, which we just remark upon in passing, is that the Taylor series of the circular and hyperbolic functions again endorse the close connections between them. The second is that all the above 'sets of convergence' are intervals and what's more 0 is always the mid-point of those intervals. We now pause to prove that result for any power series: now that we've reached the heart of this introduction to analysis you will see why we needed all our groundwork on bounds, sequences and series.

Theorem   Consider any power series

$$a_0 + a_1 x + a_2 x^2 + a_3 x^3 + \cdots$$

If the series converges for $x = x_0$ then it is absolutely convergent for any value of x closer to 0.

It follows that if D is the set of x for which the series converges then D is either $\mathbb{R}$ or is an interval of the form $(-r, r)$ or $(-r, r]$ or $[-r, r)$ or $[-r, r]$. (D is called the *inverval of convergence* and r is called the *radius of convergence* of the series, with the understanding that the series has 'infinite radius of convergence' in the case when D is $\mathbb{R}$.)

Proof Assume that the given power series converges for $x = x_0$. Then by the theorem on page 65 the individual terms of the series

$$a_0 + a_1 x_0 + a_2 x_0^2 + a_3 x_0^3 + \cdots$$

converge to 0. Hence the sequence

$$a_0, \quad a_1 x_0, \quad a_2 x_0^2, \quad a_3 x_0^3, \ldots$$

is convergent and therefore bounded (by the theorem on page 44). Let M be a number with $|a_n x_0^n| \leqslant M$ for each integer $n \geqslant 0$.

Now let x_1 be any number 'closer to 0' than x_0; i.e. x_1 is in the interval $(-|x_0|, |x_0|)$. We shall show that the series

$$a_0 + a_1 x_1 + a_2 x_1^2 + a_3 x_1^3 + \cdots$$

is absolutely convergent (and hence, by the theorem on page 70, also convergent).

Note that

$$0 \leqslant |a_n x_1^n| = |a_n x_0^n| \left|\frac{x_1}{x_0}\right|^n \leqslant M \left|\frac{x_1}{x_0}\right|^n$$

and we know that the geometric series

$$M + M \left|\frac{x_1}{x_0}\right| + M \left|\frac{x_1}{x_0}\right|^2 + M \left|\frac{x_1}{x_0}\right|^3 + \cdots$$

converges since $0 \leqslant |x_1/x_0| < 1$. Hence by the comparison test (page 66)

$$0 \leqslant |a_n x_1^n| \leqslant M \underbrace{\left|\frac{x_1}{x_0}\right|^n}_{\substack{\text{these add up} \\ \text{in series}}}$$

$$\underbrace{}_{\substack{\therefore \text{ these add} \\ \text{up in series}}}$$

Hence the series $\sum |a_n x_1^n|$ is convergent; i.e. the series

$$\sum_{n=0}^{\infty} a_n x_1^n \quad \text{or} \quad a_0 + a_1 x_1 + a_2 x_1^2 + a_3 x_1^3 + \cdots$$

is absolutely convergent (and convergent) as required.

We have therefore shown that if the power series is convergent for some number distance α from 0 then it is (absolutely) convergent for all x in the interval $(-\alpha, \alpha)$.

Let D be the set of x for which the series converges and let A be the set

$$A = \{\alpha \geqslant 0: \text{the series converges for either } x = \alpha \text{ or } x = -\alpha\}$$

Then by what we've just proved $(-\alpha, \alpha) \subseteq D$ for each $\alpha \in A$. Now *either A* is not bounded above *or* it is bounded above! In the former case there exist arbitrarily large $\alpha \in A$ and so D contains arbitrarily large intervals $(-\alpha, \alpha)$: in this case D is clearly the whole of $\mathbb{R}$. On the other hand if A *is* bounded above then (since A contains 0) it is non-empty and bounded above and so by the theorem on page 12 A has a least upper bound (or supremum) r say. We shall now show that D is one of the intervals $(-r, r)$, $(-r, r]$, $[-r, r)$ or $[-r, r]$ by showing that the set D contains the set $(-r, r)$ and is contained in the set $[-r, r]$: i.e. we show that

$$(-r, r) \subseteq D \subseteq [-r, r]$$

if $x \in (-r, r)$ then
$$|x| < r = \sup A$$
and so $|x|$ is not an upper bound of A; i.e. there exists $\alpha \in A$ with $|x| < \alpha$. Then by the above

$$x \in (-\alpha, \alpha) \subseteq D$$

this part is easy, since if $x \in D$ then $|x|$ (the distance of x from 0) is in A and so $|x| \leqslant r$ and $x \in [-r, r]$

Hence D is either $\mathbb{R}$ or is one of the stated intervals, and the theorem is proved.   □

Another property of Taylor series which became apparent in our earlier table was that if you take the series for $\sin x$ and naively 'differentiate' it term-by-term then you get the Taylor series for $\cos x$, even though there is no reason to expect this technique to make sense for infinite sums. Similarly if you differentiate the Taylor series for $\log(1 + x)$ term-by-term then you get the series for $1/(1 + x)$. It's easy to check that in general if the function f has Taylor series

$$a_0 + a_1 x + a_2 x^2 + a_3 x^3 + \cdots$$

then its derivative g $(=f')$ has Taylor series

$$a_1 + 2a_2 x + 3a_3 x^2 + 4a_4 x^3 + \cdots$$

For we know that

$$a_n = \frac{f^{(n)}(0)}{n!} \quad \left(\text{and} \quad a_{n+1} = \frac{f^{(n+1)}(0)}{(n+1)!}\right)$$

and so the coefficient of x^n in the Taylor series of g is

$$\frac{g^{(n)}(0)}{n!} = \frac{(f')^{(n)}(0)}{n!} = \frac{f^{(n+1)}(0)}{n!} = \frac{(n+1)!a_{n+1}}{n!} = (n+1)a_{n+1}$$

Hence the Taylor series for the derivative g is indeed

$$a_1 + 2a_2 x + 3a_3 x^2 + 4a_4 x^3 + \cdots$$

This new series is called the *derived* series of the original.

Note that in the example of the Taylor series of $\log(1 + x)$ and of $1/(1 + x)$ those two series have intervals of convergence $(-1, 1]$ and $(-1, 1)$ respectively: in particular they have the same radius of convergence. In fact given any power series its derived series has the same radius of convergence as the original. And we shall now see that if the series sums to $f(x)$ then f is a very nicely-behaved function within the interval of convergence and in particular its derived series sums to $f'(x)$. We have reached the limits of the sort of analysis expected in a first course and to avoid a long and technical proof only part of the theorem is proved here.

> ***Theorem***   Given a series
>
> $$a_0 + a_1 x + a_2 x^2 + a_3 x^3 + \cdots$$
>
> its derived series
>
> $$a_1 + 2a_2 x + 3a_3 x^2 + 4a_4 x^3 + \cdots$$
>
> has the same radius of convergence as the original series. Assume that the original series has interval of convergence D and define a function f by
>
> $$f(x) = a_0 + a_1 x + a_2 x^2 + a_3 x^3 + \cdots \quad x \in D$$
>
> Then f is continuous. Furthermore if the series has a positive radius of convergence r then for each $x \in (-r, r)$ f is differentiable at x and the derived series converges to $f'(x)$ (with the understanding that this works for each $x \in \mathbb{R}$ when the original series has infinite radius of convergence).

> ***Proof***   We shall simply prove the first part of the theorem. As we saw in the $\log(1 + x)$ example the original series and the derived series may have intervals of convergence which differ by the odd endpoint, but in calculating the radius of convergence such endpoints are irrelevant. So to simplify the proof slightly let us assume that the interval of convergence D has no endpoints: similarly let D' be the interval of convergence of the derived series and assume that it has no endpoints.
>
> Informally it seems that, as the coefficients of the derived series look bigger than those of the original, the new series is less likely to converge: we confirm this first by showing that $D' \subseteq D$.
>
> If $x_0 \in D'$ then there are other members of D' 'further from 0' and it follows from the previous theorem that the series
>
> $$a_1 + 2a_2 x_0 + 3a_3 x_0^2 + 4a_4 x_0^3 + \cdots$$
>
> is absolutely convergent; i.e. the series
>
> $$|a_1| + 2|a_2 x_0| + 3|a_3 x_0^2| + 4|a_4 x_0^3| + \cdots$$

is convergent. Multiplying through this by the constant $|x_0|$ we deduce that the series

$$|a_1x_0| + 2|a_2x_0{}^2| + 3|a_3x_0{}^3| + 4|a_4x_0{}^4| + \cdots$$

is also convergent. But then (by the comparison test again) the 'smaller' series

$$(|a_0| +)|a_1x_0| + |a_2x_0{}^2| + |a_3x_0{}^3| + \cdots$$

converges and, since absolute convergence implies convergence, the series

$$a_0 + a_1x_0 + a_2x_0{}^2 + a_3x_0{}^3 + \cdots$$

also converges. Hence x_0 (an arbitrary number in D') is also in D, the interval of convergence of the original series, and so we have shown that $D' \subseteq D$.

We now have to prove the rather more surprising property that $D \subseteq D'$. So assume that $x_0 \in D$: we must show that $x_0 \in D'$. This is clear if $x_0 = 0$ so assume in addition that $x_0 \neq 0$. Choose $x_1 \in D$ with $|x_1| > |x_0|$:

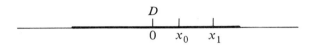

Then, as $|x_0/x_1|$ is less than 1, by the example on page 58 the sequence $(n|x_0/x_1|^n)$ converges to 0. So there is some term, the Nth say, beyond which $n|x_0/x_1|^n < |x_0|$. But then for $n \geqslant N$ we have

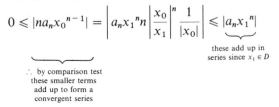

$$0 \leqslant |na_nx_0{}^{n-1}| = \left| a_nx_1{}^n n \left| \frac{x_0}{x_1} \right|^n \frac{1}{|x_0|} \right| \leqslant |a_nx_1{}^n|$$

these add up in
series since $x_1 \in D$

$\therefore$ by comparison test
these smaller terms
add up to form a
convergent series

Hence the derived series is (absolutely) convergent for $x = x_0$ and so x_0 is in D', the interval of convergence of the derived series. Therefore since x_0 was freely chosen in D it follows that $D \subseteq D'$ as required.

We have therefore shown that $D \subseteq D'$ and that $D' \subseteq D$: thus $D = D'$ and (apart from any possible endpoints) the two series have the same interval of convergence, and hence the same radius of convergence as claimed. $\square$

It follows easily from the theorem that f is differentiable arbitrarily often in $(-r, r)$ and that $f^{(n)}(0) = n!a_n$ and hence that the power series is in fact the Taylor series of f. We were led into the general idea of power series by looking specifically at Taylor series (about $x = 0$) but it turns out that any power series with a positive radius of convergence *is* a Taylor series of some function, so to some extent the two concepts coincide.

That last theorem shows us that Taylor series behave in a very natural way. For example we have seen that the Taylor series for e^x converges for all x to give

$$e^x = 1 + x + \frac{x^2}{2!} + \frac{x^3}{3!} + \cdots \quad x \in \mathbb{R}$$

Hence replacing x by $-2x$ gives

$$e^{-2x} = 1 - 2x + 2x^2 + \tfrac{4}{3}x^3 + \cdots \quad x \in \mathbb{R}$$

By the above comments this *is* the Taylor series for e^{-2x}, confirming what we saw in the example on page 180. In that sort of sensible way Taylor series can be manipulated and we shall see some further examples in the next exercises.

We have come quite a long way from elementary school differentiation and after one more set of exercises we shall be ready for the final chapter on the related (?) topic of integration.

Exercises

1   By using various tests of convergence of series established in chapter 2 find the radius of convergence of each of the following series:

(i) $1 + x^2 + \dfrac{x^4}{2!} + \dfrac{x^6}{3!} + \cdots$

(ii) $x + 2x^2 + 3x^3 + 4x^4 + \cdots$

(iii) $1 + x + 2!x^2 + 3!x^3 + \cdots$

In the case of each series with a positive radius of convergence find a neat expression for the function to which it converges (and whose Taylor series it therefore is).

2   By repeatedly differentiating the following functions find the first few terms of their Taylor series (about $x = 0$);

$$f(x) = \tan x \qquad g(x) = \tan^{-1} x$$

(In the case of f the derivatives soon get a bit complicated so you might prefer to prove first that

$$f'(x) = 1 + (f(x))^2$$

and to use this result to help you find the higher derivatives.)

Now assume that Taylor series behave in a 'natural' way and

(i) calculate the first few terms of the product of the Taylor series of tan x and the Taylor series of cos x and observe that you get the terms of the Taylor series of sin x;

(ii) use the above series for f and g to calculate the first few terms of the series of $f(g(x))$ and of the series of $g(f(x))$.

3   Differentiate the series in exercise 2 to

(i) find the first few terms of the Taylor series of sec^2 x (where sec x = $1/\cos x$);

(ii) confirm that the Taylor series of g' seems to be the series of $(1 + x^2)^{-1}$.

(Solutions on page 275)

5

An integrated conclusion

A familiar area

We've already touched upon the idea of the area under the graph of a function f from $x = a$ to $x = b$. Indeed we used the idea to define the log function. But can we make this idea work for any function?

Example Let f be the function given by

$$f(x) = x^2 \qquad x \in \mathbb{R}$$

From basic principles find under- and over-estimates of the area under the graph of f from $x = 0$ to $x = 1$. Hence find the area.

Solution Let the required area be A. We can divide the region under the graph of f between $x = 0$ and $x = 1$ into vertical strips, and on each strip try to fit as tall a rectangle as possible under the graph of f, as illustrated for five strips on the left below. To make the calculations easy in this case we've made the strips of equal width. Now the total area of those

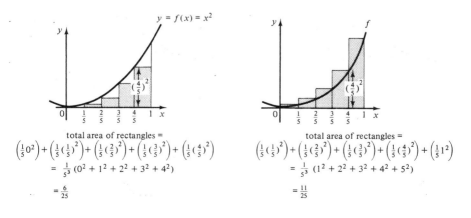

total area of rectangles $=$

$$\left(\tfrac{1}{5}0^2\right) + \left(\tfrac{1}{5}\left(\tfrac{1}{5}\right)^2\right) + \left(\tfrac{1}{5}\left(\tfrac{2}{5}\right)^2\right) + \left(\tfrac{1}{5}\left(\tfrac{3}{5}\right)^2\right) + \left(\tfrac{1}{5}\left(\tfrac{4}{5}\right)^2\right)$$

$$= \tfrac{1}{5^3}\left(0^2 + 1^2 + 2^2 + 3^2 + 4^2\right)$$

$$= \tfrac{6}{25}$$

total area of rectangles $=$

$$\left(\tfrac{1}{5}\left(\tfrac{1}{5}\right)^2\right) + \left(\tfrac{1}{5}\left(\tfrac{2}{5}\right)^2\right) + \left(\tfrac{1}{5}\left(\tfrac{3}{5}\right)^2\right) + \left(\tfrac{1}{5}\left(\tfrac{4}{5}\right)^2\right) + \left(\tfrac{1}{5}1^2\right)$$

$$= \tfrac{1}{5^3}\left(1^2 + 2^2 + 3^2 + 4^2 + 5^2\right)$$

$$= \tfrac{11}{25}$$

rectangles gives an under-estimate of A. Similarly, on each strip we can draw a rectangle which just covers f, as shown on the right, and the total area of those rectangles gives an over-estimate of A.

Hence A is sandwiched between the two estimates

$$\tfrac{6}{25} \leqslant A \leqslant \tfrac{11}{25}$$

If we now increase the number of strips to 10 and repeat the procedure we'll obtain closer under- and over-estimates of A as shown below

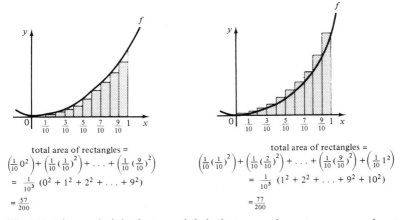

total area of rectangles =
$$\left(\tfrac{1}{10}0^2\right)+\left(\tfrac{1}{10}(\tfrac{1}{10})^2\right)+\ldots+\left(\tfrac{1}{10}(\tfrac{9}{10})^2\right)$$

$$= \tfrac{1}{10^3}(0^2+1^2+2^2+\ldots+9^2)$$

$$= \tfrac{57}{200}$$

total area of rectangles =
$$\left(\tfrac{1}{10}(\tfrac{1}{10})^2\right)+\left(\tfrac{1}{10}(\tfrac{2}{10})^2\right)+\ldots+\left(\tfrac{1}{10}(\tfrac{9}{10})^2\right)+\left(\tfrac{1}{10}1^2\right)$$

$$= \tfrac{1}{10^3}(1^2+2^2+\ldots+9^2+10^2)$$

$$= \tfrac{77}{200}$$

Hence A is sandwiched more tightly between these two new estimates

$$\tfrac{57}{200} \leqslant A \leqslant \tfrac{77}{200}$$

In general if we divide the region into n strips of equal width the under- and over-estimates obtained for A are as follows (where we use the formula for the sum of consecutive squares as derived in exercise 5 on page 19):

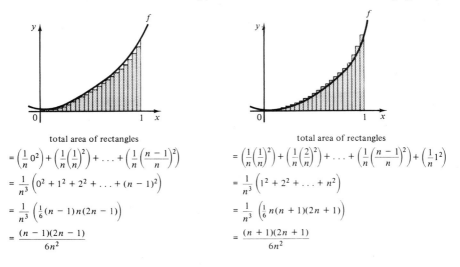

total area of rectangles
$$= \left(\tfrac{1}{n}0^2\right)+\left(\tfrac{1}{n}\left(\tfrac{1}{n}\right)^2\right)+\ldots+\left(\tfrac{1}{n}\left(\tfrac{n-1}{n}\right)^2\right)$$

$$= \tfrac{1}{n^3}\left(0^2+1^2+2^2+\ldots+(n-1)^2\right)$$

$$= \tfrac{1}{n^3}\left(\tfrac{1}{6}(n-1)n(2n-1)\right)$$

$$= \tfrac{(n-1)(2n-1)}{6n^2}$$

total area of rectangles
$$= \left(\tfrac{1}{n}\left(\tfrac{1}{n}\right)^2\right)+\left(\tfrac{1}{n}\left(\tfrac{2}{n}\right)^2\right)+\ldots+\left(\tfrac{1}{n}\left(\tfrac{n-1}{n}\right)^2\right)+\left(\tfrac{1}{n}1^2\right)$$

$$= \tfrac{1}{n^3}\left(1^2+2^2+\ldots+n^2\right)$$

$$= \tfrac{1}{n^3}\left(\tfrac{1}{6}n(n+1)(2n+1)\right)$$

$$= \tfrac{(n+1)(2n+1)}{6n^2}$$

Hence A is sandwiched between these two estimates and for each positive integer n we have

$$\frac{(n-1)(2n-1)}{6n^2} \leqslant A \leqslant \frac{(n+1)(2n+1)}{6n^2}$$

What happens for larger and larger n? As you might expect the estimates get closer and closer and we have

$$\frac{(n-1)(2n-1)}{6n^2} \leqslant A \leqslant \frac{(n+1)(2n+1)}{6n^2}$$
$$\downarrow \qquad\qquad \Vert \qquad\qquad \downarrow$$
$$\tfrac{1}{3} \qquad \therefore\quad \Vert \qquad\qquad \tfrac{1}{3}$$
$$\tfrac{1}{3}$$

So it seems that the area A is $\tfrac{1}{3}$.   □

How can we describe that process formally and adapt it to find the area under the graph of any function f between $x = a$ and $x = b$? We wish first to choose some numbers between a and b (these are going to be the places where we draw our vertical lines). Given any finite subset P of $[a, b]$ which includes the numbers a and b we can put the members of P into increasing order and call them

$$a = a_0 < a_1 < a_2 < \cdots < a_{n-1} < a_n = b$$

say: the set P is called a *partition* of $[a, b]$. Given such a partition we can then divide the coordinate plane between $x = a$ and $x = b$ into vertical strips

$$a_0 \leqslant x \leqslant a_1, \quad a_1 \leqslant x \leqslant a_2, \quad \ldots \quad, a_{n-1} \leqslant x \leqslant a_n$$

as illustrated below.

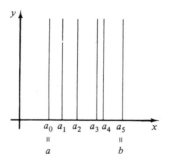

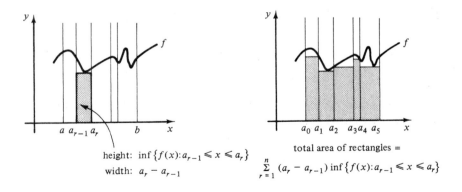

height: $\inf\{f(x): a_{r-1} \leqslant x \leqslant a_r\}$

width: $a_r - a_{r-1}$

total area of rectangles =

$$\sum_{r=1}^{n} (a_r - a_{r-1}) \inf\{f(x): a_{r-1} \leqslant x \leqslant a_r\}$$

If we now consider the graph of a function f like the one shown on the left above, what is the biggest rectangle that we can fit under the graph in the strip $a_{r-1} \leqslant x \leqslant a_r$? It seems that the height of the rectangle is the minimum value of the function f for $a_{r-1} \leqslant x \leqslant a_r$. But our picture is of a nice continuous function which actually has a minimum there: in general the height must be the biggest number which is less than or equal to the value of the function on that strip; i.e. the greatest lower bound (or infimum) of the set

$$\{f(x): a_{r-1} \leqslant x \leqslant a_r\}$$

The rectangle will then have area

$$(a_r - a_{r-1}) \times \inf\{f(x): a_{r-1} \leqslant x \leqslant a_r\}$$

So for the partition P the total area of such rectangles under f's graph is

$$\sum_{r=1}^{n} (a_r - a_{r-1}) \times \inf\{f(x): a_{r-1} \leqslant x \leqslant a_r\}$$

and that number will give an under-estimate of the required area under the graph of f (as illustrated in the right-hand figure above).

Let L be the set of all the under-estimates of the area obtained in that way from all the different partitions of $[a, b]$ (**L**ess than the required area).

Similarly we can obtain an over-estimate of the area from each partition $a = a_0 < a_1 < a_2 < \cdots < a_{n-1} < a_n = b$ of $[a, b]$. This time the height of the rectangle on the strip $a_{r-1} \leqslant x \leqslant a_r$ is the smallest number which is greater than or equal to the values the function takes there; i.e. it is the least upper bound (or supremum) of the set

$$\{f(x): a_{r-1} \leqslant x \leqslant a_r\}$$

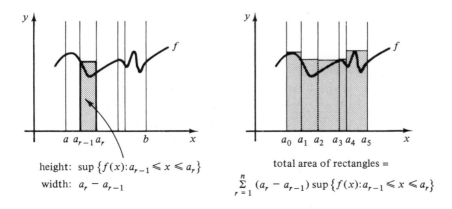

height: $\sup\{f(x): a_{r-1} \leqslant x \leqslant a_r\}$

width: $a_r - a_{r-1}$

total area of rectangles =

$\sum\limits_{r=1}^{n} (a_r - a_{r-1}) \sup\{f(x): a_{r-1} \leqslant x \leqslant a_r\}$

Hence that rectangle has area

$$(a_r - a_{r-1}) \times \sup\{f(x): a_{r-1} \leqslant x \leqslant a_r\}$$

and the total area of the rectangles obtained in this way is

$$\sum_{r=1}^{n} (a_r - a_{r-1}) \times \sup\{f(x): a_{r-1} \leqslant x \leqslant a_r\}$$

Let M be the set of all the over-estimates of the area obtained in that way (**M**ore than the required area).

Having defined the sets L and M of all the under- and over-estimates of the required area under the graph of f, let us recall how we used some of those estimates in the previous worked example. First we sandwiched the area A as follows:

$$\underset{\in L}{\underbrace{\tfrac{6}{25}}} \leqslant A \leqslant \underset{\in M}{\underbrace{\tfrac{11}{25}}}$$

Then we found closer estimates and sandwiched A as follows:

$$\underset{\in L}{\underbrace{\tfrac{57}{200}}} \leqslant A \leqslant \underset{\in M}{\underbrace{\tfrac{77}{200}}}$$

And in general for each positive integer n we sandwiched A as follows:

$$\overbrace{\frac{(n-1)(2n-1)}{6n^2}}^{\in L} \leqslant A \leqslant \overbrace{\frac{(n+1)(2n+1)}{6n^2}}^{\in M}$$

$$\downarrow \qquad\qquad\qquad \downarrow$$

$$\tfrac{1}{3} \qquad\qquad\qquad \tfrac{1}{3}$$

In effect we had found some partitions which led to a sequence in L and to a sequence in M with the common limit of $\frac{1}{3}$, which is what the area turned out to be.

We can now use our ideas about L and M to define what we mean by the 'area under a graph' and more generally by the notion of an 'integral' as developed by the German mathematician Georg Riemann in the mid-nineteenth century. For the purposes of illustration we only considered the 'area under the graph' of a function whose graph was above the x-axis but we now drop that restriction and extend the idea from area to that of an 'integral' (where regions under the x-axis will actually give negative answers). However, since the definition of the members of L and M will involve 'infs' and 'sups' of the function f, we have to restrict attention to intervals where f is bounded so that these terms are well defined.

> **Definition**   Let f be a function whose domain includes the interval $[a, b]$ and which is bounded on $[a, b]$. For each partition $\{a_0, a_1, a_2, \ldots, a_n\}$ of $[a, b]$ with
>
> $$a = a_0 < a_1 < a_2 < \cdots < a_n = b$$
>
> let l and m be given by
>
> $$l = \sum_{r=1}^{n} (a_r - a_{r-1}) \times \inf\{f(x): a_{r-1} \leqslant x \leqslant a_r\}$$
>
> $$m = \sum_{r=1}^{n} (a_r - a_{r-1}) \times \sup\{f(x): a_{r-1} \leqslant x \leqslant a_r\}$$
>
> Then let L be the set of numbers l arising in that way from all the partitions of $[a, b]$ and similarly let M be the set of all numbers m arising in that way. If there are some partitions which lead to both
>
> $$\underbrace{l_1, l_2, l_3, \ldots}_{\in L} \quad \text{and} \quad \underbrace{m_1, m_2, m_3, \ldots}_{\in M}$$
>
> such that the two sequences have the same limit α then f is said to be *integrable over* $[a, b]$ and α is called its *integral*. We write
>
> $$\int_a^b f = \alpha \quad \text{or} \quad \int_a^b f(x)\, \mathrm{d}x = \alpha$$

Before proceeding to use that definition there are some comments which should be made. The first concerns the notation: in fact

$$\int_a^b f$$

is quite sufficient – it means the integral of f between a and b. The more traditional notation includes the variable:

$$\int_a^b f(x)\,dx$$

This can be read as 'the integral of $f(x)$ with respect to x between $x = a$ and $x = b$'. In many ways the x is redundant and

$$\int_a^b f(t)\,dt$$

gives precisely the same numerical answer. The traditional notation is motivated by the fact that the integral is introduced by adding up lots of rectangles of height $f(x)$ of width δx where (as when we justified the notation dy/dx) δx represents a small change in x. The limiting process described in the definition is shown in the notation by changing

$$\sum f(x) \times \delta x \quad \text{to} \quad \int f(x)\,dx$$

It turns out (as we shall see later) that this fits in quite neatly with the dy/dx notation for a derivative.

Our second comment is that the definition of the integral assumes that $a \leqslant b$. In fact if $a = b$ then the definition becomes trivial: the only partition of $[a, a]$ is the set consisting of the single point a and that gives both under- and over-estimate of the integral as 0. Hence any function whose domain includes a is integrable over $[a, a]$ with

$$\int_a^a f = 0$$

We shall see later that integrals behave in quite a neat algebraic way and we shall need the idea of

$$\int_a^b f$$

even when $a > b$. We simply define that integral by

$$\int_a^b f = -\int_b^a f$$

provided that f is integrable over $[b, a]$.

We can now proceed to a worked example of an integral obtained from the basic principles outlined in the definition.

Example   Show that the function f given by

$$f(x) = x \qquad x \in \mathbb{R}$$

is integrable over [0, 1], and find its integral

$$\int_0^1 f \quad \text{or} \quad \int_0^1 f(x)\,dx \quad \text{or} \quad \int_0^1 x\,dx$$

Solution   To show that f is integrable we have to find some partition such that their under- and over-estimates

$$\underbrace{l_1, l_2, l_3, \ldots}_{\in L} \quad \text{and} \quad \underbrace{m_1, m_2, m_3, \ldots}_{\in M}$$

have a common limit, and that limit is then the integral. As is quite common with reasonably well-behaved functions the partitions which give strips of equal width serve our purpose. For example the partition $\{0, \frac{1}{5}, \frac{2}{5}, \frac{3}{5}, \frac{4}{5}, 1\}$ of [0, 1] gives

$$l = \sum_{r=1}^{5} (a_r - a_{r-1})$$
$$\times \inf\{f(x): a_{r-1} \leqslant x \leqslant a_r\}$$
$$= \sum_{r=1}^{5} \left(\frac{r}{5} - \frac{r-1}{5}\right)$$
$$\times \underbrace{\inf\left\{x: \frac{r-1}{5} \leqslant x \leqslant \frac{r}{5}\right\}}_{= \frac{r-1}{5}}$$
$$= (\tfrac{1}{5} \times 0) + (\tfrac{1}{5} \times \tfrac{1}{5}) + (\tfrac{1}{5} \times \tfrac{2}{5})$$
$$+ (\tfrac{1}{5} \times \tfrac{3}{5}) + (\tfrac{1}{5} \times \tfrac{4}{5})$$
$$= \frac{1}{5^2}(0 + 1 + 2 + 3 + 4)$$
$$= \tfrac{2}{5}$$

as illustrated

$$m = \sum_{r=1}^{5} (a_r - a_{r-1})$$
$$\times \sup\{f(x): a_{r-1} \leqslant x \leqslant a_r\}$$
$$= \sum_{r=1}^{5} \left(\frac{r}{5} - \frac{r-1}{5}\right)$$
$$\times \underbrace{\sup\left\{x: \frac{r-1}{5} \leqslant x \leqslant \frac{r}{5}\right\}}_{= \frac{r}{5}}$$
$$= (\tfrac{1}{5} \times \tfrac{1}{5}) + (\tfrac{1}{5} \times \tfrac{2}{5}) + (\tfrac{1}{5} \times \tfrac{3}{5})$$
$$+ (\tfrac{1}{5} \times \tfrac{4}{5}) + (\tfrac{1}{5} \times 1)$$
$$= \frac{1}{5^2}(1 + 2 + 3 + 4 + 5)$$
$$= \tfrac{3}{5}$$

as illustrated

$y = f(x) = x$

In general we'll consider the partition $\{0, 1/n, 2/n, \ldots, 1\}$ which gives n strips of equal width and leads to

$$l_n = \sum_{r=1}^{n} (a_r - a_{r-1})$$

$$\times \inf\{f(x): a_{r-1} \leqslant x \leqslant a_r\}$$

$$= \sum_{r=1}^{n} \left(\frac{r}{n} - \frac{r-1}{n}\right)$$

$$\times \underbrace{\inf\left\{x: \frac{r-1}{n} \leqslant x \leqslant \frac{r}{n}\right\}}_{= \dfrac{r-1}{n}}$$

$$= \left(\frac{1}{n} \times 0\right) + \left(\frac{1}{n} \times \frac{1}{n}\right) + \left(\frac{1}{n} \times \frac{2}{n}\right)$$

$$+ \cdots + \left(\frac{1}{n} \times \frac{n-1}{n}\right)$$

$$= \frac{1}{n^2}(0 + 1 + 2 + \cdots + (n-1))$$

$$= \frac{1}{n^2}(\tfrac{1}{2}(n-1)n) = \frac{n-1}{2n}$$

So the set L of under-estimates includes the numbers $(n-1)/2n$ for each positive integer n. So L includes the sequence

$$0, \tfrac{1}{4}, \tfrac{1}{3}, \tfrac{3}{8}, \tfrac{2}{5}, \tfrac{5}{12}, \tfrac{3}{7}, \tfrac{7}{16}, \ldots$$

which converges to $\tfrac{1}{2}$.

$$m_n = \sum_{r=1}^{n} (a_r - a_{r-1})$$

$$\times \sup\{f(x): a_{r-1} \leqslant x \leqslant a_r\}$$

$$= \sum_{r=1}^{n} \left(\frac{r}{n} - \frac{r-1}{n}\right)$$

$$\times \underbrace{\sup\left\{x: \frac{r-1}{n} \leqslant x \leqslant \frac{r}{n}\right\}}_{= \dfrac{r}{n}}$$

$$= \left(\frac{1}{n} \times \frac{1}{n}\right) + \left(\frac{1}{n} \times \frac{2}{n}\right) + \left(\frac{1}{n} \times \frac{3}{n}\right)$$

$$+ \cdots + \left(\frac{1}{n} \times 1\right)$$

$$= \frac{1}{n^2}(1 + 2 + 3 + \cdots + n)$$

$$= \frac{1}{n^2}(\tfrac{1}{2}n(n+1)) = \frac{n+1}{2n}$$

So the set M of over-estimates includes the numbers $(n+1)/2n$ for each positive integer n. So M includes the sequence

$$1, \tfrac{3}{4}, \tfrac{2}{3}, \tfrac{5}{8}, \tfrac{3}{5}, \tfrac{7}{12}, \tfrac{4}{7}, \tfrac{9}{16}, \ldots$$

which converges to $\tfrac{1}{2}$.

We have therefore found some partitions leading to sequences $l_1, l_2, l_3, \ldots \in L$ and $m_1, m_2, m_3, \ldots \in M$ with a common limit, as required. It follows that f is integrable over $[0, 1]$ and that its integral is $\tfrac{1}{2}$; i.e.

$$\int_0^1 f = \tfrac{1}{2} \quad \text{or} \quad \int_0^1 f(x)\,dx = \tfrac{1}{2} \quad \text{or} \quad \int_0^1 x\,dx = \tfrac{1}{2}$$

Does $\tfrac{1}{2}$ make a sensible answer for the area under the graph of f between $x = 0$ and $x = 1$? Of course it does because that area is just a triangle of base 1 and height 1.

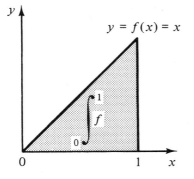

No-one in their right minds would use this long-winded method for finding that area but it does show that our new idea of integral seems to lead to sensible answers.   □

We tentatively assumed in the definition and in the previous discussion that every number in L is less than (or perhaps equal to) every number in M and that the required area was 'sandwiched between' L and M, i.e. greater than or equal to each member of L and less than or equal to each member of M. (This is reminiscent of the way we introduced the completeness axiom on page 10 and the intermediate value theorem on pages 133–9.) The definition of an integral requires us to find some partitions leading to both under-estimates $l_1, l_2, l_3, \ldots \in L$ and to over-estimates $m_1, m_2, m_3, \ldots \in M$ where both these sequences converge to the same limit, α say. This number will then turn out to be the unique 'piggy-in-the-middle' between L and M, as we now see.

Theorem   Let f be a function whose domain includes the interval $[a, b]$ and which is bounded on $[a, b]$. Let L and M be the sets of numbers constructed as in the definition above. Then:

(i) $l \leqslant m$ for each $l \in L$ and $m \in M$;
(ii) there will be a unique number sandwiched between L and M if and only if there are some partitions which lead simultaneously to sequences $l_1, l_2, l_3, \ldots$ in L and $m_1, m_2, m_3, \ldots$ in M both of which converge to the same limit (i.e. if and only if f is integrable over $[a, b]$). That common limit (which is the integral) will then equal the unique number between L and M.

Proof   We make two key observations before proving the

theorem. The first is that for any partition P the 'under-estimate' $l \in L$ arising from P is less than or equal to the 'over-estimate' $m \in M$ arising from P. This is because the infimum of any set is clearly less than or equal to the supremum of the set and so

$$l = \sum_{r=1}^{n} (a_r - a_{r-1}) \times \inf\{f(x): a_{r-1} \leqslant x \leqslant a_r)$$

$$\leqslant \sum_{r=1}^{n} (a_r - a_{r-1}) \times \sup\{f(x): a_{r-1} \leqslant x \leqslant a_r\} = m$$

The second key point is that if a partition P leads to under- and over-estimates l and m then adding a new number to P will lead to under- and over-estimates l' and m' with $l \leqslant l'$ and $m \geqslant m'$. This is illustrated below:

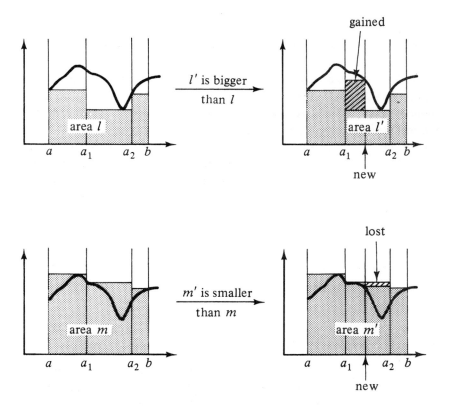

Suppose that the extra number added to P is a' and that it lies between the numbers a_{r-1} and a_r in P. Then the increase in l occurs because the single rectangle on the strip $a_{r-1} \leqslant x \leqslant a_r$ is replaced by two, one on

$a_{r-1} \leqslant x \leqslant a'$ and the other on $a' \leqslant x \leqslant a_r$ with

area of the two new rectangles =

$$(a' - a_{r-1}) \times \underbrace{\inf\{f(x): a_{r-1} \leqslant x \leqslant a'\}}_{\substack{\text{this is the inf of} \\ \text{a subset and} \\ \text{so it's the same} \\ \text{or bigger than} \\ \inf\{f(x): a_{r-1} \leqslant x \leqslant a_r\}}} + (a_r - a') \times \underbrace{\inf\{f(x): a' \leqslant x \leqslant a_r\}}_{\substack{\text{this is the inf of} \\ \text{a subset and} \\ \text{so it's the same} \\ \text{or bigger than} \\ \inf\{f(x): a_{r-1} \leqslant x \leqslant a_r\}}}$$

$\geqslant (a' - a_{r-1}) \inf\{f(x): a_{r-1} \leqslant x \leqslant a_r\} + (a_r - a') \inf\{f(x): a_{r-1} \leqslant x \leqslant a_r\}$

$= (a_r - a_{r-1}) \times \inf\{f(x): a_{r-1} \leqslant x \leqslant a_r\}$

= area of the original single rectangle

So adding extra numbers to the partition will increase the under-estimates (and similarly decrease the over-estimates). (That's not surprising – the more numbers you take in the partition the closer you'd expect the two estimates to be.)

Having established those key points we are now ready to prove the theorem.

(i) Let $l \in L$ be an under-estimate arising from the partition P_1 of $[a, b]$ and let $m \in M$ be an over-estimate arising from the partition P_2 of $[a, b]$. Then let $P = P_1 \cup P_2$; i.e. P consists of all the numbers in either P_1 or P_2 (or both). It follows that P consists of P_1 with possibly some numbers added *and* that P consists of P_2 with possibly some numbers added. So now let the partition P lead to an under-estimate l' and an over-estimate m'. By the two key facts established above we have

$$l' \leqslant m'$$

and

$$l \leqslant l' \quad \text{and} \quad m' \leqslant m$$

It follows that

$$l \leqslant l' \leqslant m' \leqslant m$$

and that $l \leqslant m$ as required.

(ii) We have now shown that $l \leqslant m$ for each $l \in L$ and $m \in M$, and we have two sets of the type we considered when introducing the completeness axiom. By that axiom there exists at least one number between L and M. We shall show that these sets have a **unique** 'piggy-in-the-middle' α **if and only if** there exists partitions of $[a, b]$ which lead simultaneously to sequences

$$\underbrace{l_1, l_2, l_3, \ldots \to \alpha}_{\in L} \quad \text{and} \quad \underbrace{m_1, m_2, m_3, \ldots \to \alpha}_{\in M}$$

So assume first that L and M do have a unique number α 'between' them. Then (as in exercise 3 on page 13)

$$\alpha = \sup L = \inf M$$

Since α is the least upper bound of L it follows that $\alpha - 1$ is not an upper bound of L and that some partition P_1 leads to an under-estimate $l \in L$ with $l > \alpha - 1$. Similarly $\alpha + 1$ is not a lower bound of M and there is a partition P_2 which leads to an over-estimate $m \in M$ with $m < \alpha + 1$. By our earlier comments the partition $P_1 \cup P_2$ leads to under- and over-estimates l_1 and m_1 with

$$\alpha - 1 < l \leqslant l_1 \leqslant m_1 \leqslant m < \alpha + 1$$

Similarly there is a partition of $[a, b]$ which gives under- and over-estimates l_2 and m_2 with

$$\alpha - \tfrac{1}{2} < l_2 \leqslant m_2 < \alpha + \tfrac{1}{2}$$

and there exists a partition which gives

$$\alpha - \tfrac{1}{3} < l_3 \leqslant m_3 < \alpha + \tfrac{1}{3}$$
$$\vdots$$

Hence there exist partitions leading simultaneously to

$$\underbrace{l_1, l_2, l_3, \ldots \to \alpha}_{\in L} \quad \text{and} \quad \underbrace{m_1, m_2, m_3, \ldots \to \alpha}_{\in M}$$

as required.

Conversely assume that such sequences exist; i.e.

$$\underbrace{l_1, l_2, l_3, \ldots \to \alpha}_{\in L} \quad \text{and} \quad \underbrace{m_1, m_2, m_3, \ldots \to \alpha}_{\in M}$$

We know by the completeness axiom that there is at least one number 'between' L and M. For any such number β we have

$$l \leqslant \beta \leqslant m$$

for each $l \in L$ and $m \in M$. In particular

$$l_n \leqslant \beta \leqslant m_n$$

for each positive integer n. Letting $n \to \infty$ we have

$$
\begin{array}{ccc}
l_n & \leqslant \beta \leqslant & m_n \\
\downarrow & \| & \downarrow \\
\alpha \;\therefore & & \alpha \\
& \alpha &
\end{array}
$$

and so $\beta = \alpha$. Hence α is the *only* number between L and M, and the theorem is finally proved.   □

It is worth making a few comments about L and M for future reference. Firstly note that we restricted the definition and theorem to a function f bounded on $[a, b]$ so that the sets L and M (whose members were obtained by taking 'infs' and 'sups' of f) would be sensibly defined. For to say that f is bounded on $[a, b]$ means that the (non-empty) set

$$\{f(x): a \leqslant x \leqslant b\}$$

is bounded and that it has an infimum and a supremum

$$\inf\{f(x): a \leqslant x \leqslant b\} \quad \text{and} \quad \sup\{f(x): a \leqslant x \leqslant b\}$$

Now if we consider the trivial partition of $[a, b]$ consisting of just the numbers a and b then we obtain the following under- and over-estimates of the integral of f:

$$(b - a) \inf\{f(x): a \leqslant x \leqslant b\} \in L \quad \text{and}$$
$$(b - a) \sup\{f(x): a \leqslant x \leqslant b\} \in M$$

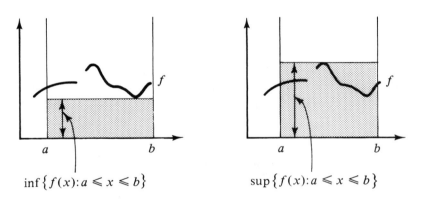

$$\inf\{f(x): a \leqslant x \leqslant b\} \qquad\qquad \sup\{f(x): a \leqslant x \leqslant b\}$$

Hence L and M are non-empty and if f is integrable over $[a, b]$ then its integral satisfies

$$(b - a) \inf\{f(x): a \leqslant x \leqslant b\} \quad \leqslant \quad \int_a^b f \quad \leqslant \quad (b - a) \sup\{f(x): a \leqslant x \leqslant b\}$$

By the theorem $l \leqslant m$ for each $l \in L$ and $m \in M$. It follows that the non-empty set L is bounded above and has a supremum and the non-empty set M is bounded below and has an infimum and that

$$\sup L \leqslant \inf M$$

In fact, as we observed in the proof of the theorem, f is integrable over $[a, b]$ if and only if there is a unique number 'between' L and M and this happens if and only if

$$\sup L = \inf M$$

The integral will then be that common value.

Putting together all these facts we get

$$(b-a)\inf\{f(x): a \leqslant x \leqslant b\} \leqslant \sup L \leqslant \inf M \leqslant (b-a)\sup\{f(x): a \leqslant x \leqslant b\}$$

$$\underbrace{\phantom{(b-a)\inf\{f(x): a \leqslant x \leqslant b\}}}_{\in L} \qquad \uparrow \qquad \underbrace{\phantom{(b-a)\sup\{f(x): a \leqslant x \leqslant b\}}}_{\in M}$$

with equality if
and only if f is
integrable (and the
common inf/sup *is*
the integral)

Aren't we making rather heavy weather of this? Wouldn't approximating to the area with just under-estimates have been enough? Won't all functions be integrable? No.

> **Example** Let f be the function given by
> $$f(x) = \begin{cases} 1 & \text{if } x \text{ is irrational} \\ 0 & \text{if } x \text{ is rational} \end{cases}$$
> and let a, b be numbers with $a < b$. Show that f is not integrable over $[a, b]$.

> **Solution** Recall (from exercise 4 on page 18) that between any two numbers there are both rationals and irrationals. Hence for any partition $a = a_0 < a_1 < a_2 < \cdots a_n = b$ of $[a, b]$ there exist rationals and irrationals in the set $\{x: a_{r-1} \leqslant x \leqslant a_r\}$. It follows that the set $\{f(x): a_{r-1} \leqslant x \leqslant a_r\}$ is the set $\{0, 1\}$. Therefore for that partition
> $$l = \sum_{r=1}^{n} (a_r - a_{r-1}) \underbrace{\inf\{f(x): a_{r-1} \leqslant x \leqslant a_r\}}_{=\inf\{0, 1\} = 0} = \sum_{r=1}^{n} (a_r - a_{r-1}) \times 0 = 0$$
> and
> $$m = \sum_{r=1}^{n} (a_r - a_{r-1}) \underbrace{\sup\{f(x): a_{r-1} \leqslant x \leqslant a_r\}}_{=\sup\{0, 1\} = 1} = \sum_{r=1}^{n} (a_r - a_{r-1}) \times 1$$
> $$= \left\{ \begin{array}{l} (a_1 - a) + \\ (a_2 - a_1) + \\ (a_3 - a_2) + \\ \quad \vdots \\ (b - a_{n-1}) \end{array} \right\} = b - a$$

So in this case *every* partition gives $l = 0$ and $m = b - a$. Therefore the sets L and M contain just one number each,

$$L = \{0\} \quad \text{and} \quad M = \{b - a\}$$

and since $b - a > 0$ there are no sequences in L and in M with a common limit. (Alternatively note that there are *many* numbers 'between' L and M.)

It follows from the definition (or from the above theorem) that f is not integrable over $[a, b]$.   □

Those examples were quite hard work but luckily, as with most of the other analytical topics which we've met, we shall soon find some much quicker techniques and (after the exercises) we shall rarely have to resort to the definition.

Exercises

1   (i) For each non-negative integer r let t_r be the *triangular number* $\frac{1}{2}r(r + 1)$. Show that

$$t_r^2 - t_{r-1}^2 = r^3$$

and deduce that

$$1^3 + 2^3 + 3^3 + \cdots + n^3 = (\tfrac{1}{2}n(n + 1))^2$$

(ii) Let f be the function given by

$$f(x) = x^3 \quad x \in \mathbb{R}$$

Show that f is integrable over $[0, 1]$ and find

$$\int_0^1 x^3 \, dx$$

2   (i) Let α be a number which is not a multiple of 2π and let r be a positive integer. Show that

$$\sin(r\alpha) = \frac{\cos((r - \tfrac{1}{2})\alpha) - \cos((r + \tfrac{1}{2})\alpha)}{2 \sin \tfrac{1}{2}\alpha}$$

and deduce that

$$\sin \alpha + \sin 2\alpha + \sin 3\alpha + \cdots + \sin n\alpha = \frac{\cos \tfrac{1}{2}\alpha - \cos((n + \tfrac{1}{2})\alpha)}{2 \sin \tfrac{1}{2}\alpha}$$

(ii) Let f be the function given by

$$f(x) = \sin x \quad x \in \mathbb{R}$$

Show that f is integrable over $[0, \pi/2]$ and find

$$\int_0^{\pi/2} \sin x \, dx$$

3   Let $a < b$ and let f be an increasing function which is bounded on $[a, b]$. Consider the partition $a = a_0 < a_1 < \cdots < a_n = b$ which divides $[a, b]$ into n equal strips and let

$$l_n = \sum_{r=1}^{n} (a_r - a_{r-1}) \inf\{f(x) \colon a_{r-1} \leqslant x \leqslant a_r\}$$

$$m_n = \sum_{r=1}^{n} (a_r - a_{r-1}) \sup\{f(x) \colon a_{r-1} \leqslant x \leqslant a_r\}$$

Show that $m_n - l_n \to 0$ as $n \to \infty$ and deduce that f is integrable over $[a, b]$.

Prove that the function f given by

$$f(x) = [x^3] \quad \text{(the 'integer part' of } x^3) \qquad x \in \mathbb{R}$$

is integrable over any interval $[a, b]$ and find

$$\int_0^{1\frac{1}{2}} [x^3] \, dx$$

4   We saw in the worked example on page 189 that

$$\int_0^1 x^2 \, dx = \tfrac{1}{3}$$

Prove that the same function is integrable over $[a, b]$ for any numbers a and b and find

$$\int_a^b x^2 \, dx$$

5   Let f be a function which is integrable over $[a, b]$ and let g be the function $-f$; i.e. $g(x) = -f(x)$ for each x. Let L and M be the usual sets of under- and over-estimates of the integral of f over $[a, b]$ and let L' and M' be the corresponding sets for g. Show that

 (i) $l \in L$ if and only if $-l \in M'$;
 (ii) $m \in M$ if and only if $-m \in L'$;
 (iii) g is integrable over $[a, b]$ with

$$\int_a^b (-f) = \int_a^b g = - \int_a^b f$$

(Solutions on page 277)

An even more familiar area

As promised we shall now begin to establish some properties of integrals which will enable us to calculate them without constantly resorting to that formidable definition.

Note firstly that the integral was designed to coincide with the 'area under the graph' of a function. But what if the function f is negative? The integral is still the area sandwiched between the graph and the x-axis, but now it is negative. That follows from the fact (established in the previous exercises) that

$$\int_a^b (-f) = - \int_a^b f$$

as we now illustrate:

for a graph above the x-axis   for a graph below the x-axis

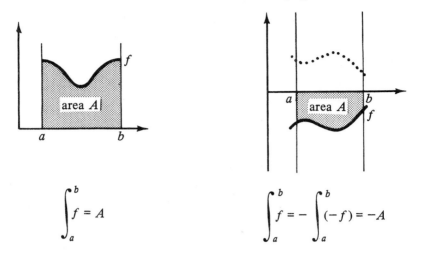

$$\int_a^b f = A \qquad\qquad \int_a^b f = -\int_a^b(-f) = -A$$

But what about functions whose graphs lie partly above and partly below the x-axis? It turns out that their integrals are the total obtained from a positive contribution for the area above the x-axis and a negative contribution for the area below the axis, e.g.

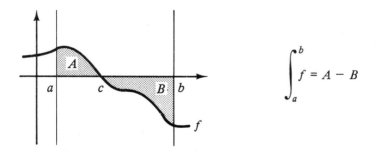

$$\int_a^b f = A - B$$

To see this we need the result that

$$\int_a^b f = \int_a^c f + \int_c^b f$$

But rather than prove that result (which would have to be restricted to functions integrable over each of those intervals) we prove a variation which holds for any function which is bounded on those intervals. This is not a diversion taken merely on a pure mathematician's whim – it will have a marvellous pay-off in a couple of pages' time.

We noted in the previous section that for a function f bounded on the interval $[a, b]$ there are well-defined sets L and M of under- and over-estimates of its integral arising from partitions of $[a, b]$. These sets have the property that

$$\sup L \leqslant \inf M$$

with equality holding if and only if f is integrable over $[a, b]$, their common value then being the integral. So now let us define the *lower* and *upper integrals* of f by

$$\int_{\underline{a}}^{b} f = \sup L \quad \text{and} \quad \overline{\int_{a}^{b}} f = \inf M$$

These are well defined for any function f which is bounded on $[a, b]$ and f will be integrable there if and only if

$$\int_{\underline{a}}^{b} f = \overline{\int_{a}^{b}} f$$

in which case their common value is the usual integral. Again we can extend these ideas to the cases where $a > b$ by defining

$$\int_{\underline{a}}^{b} f = -\int_{\underline{b}}^{a} f \quad \text{and} \quad \overline{\int_{a}^{b}} f = -\overline{\int_{b}^{a}} f$$

We shall now return to the result quoted above and prove it for the lower and upper integrals.

Theorem   Let a, b and c be any numbers and let f be a function which is bounded on some interval containing those numbers. Then

$$\int_{\underline{a}}^{b} f = \int_{\underline{a}}^{c} f + \int_{\underline{c}}^{b} f \quad \text{and} \quad \overline{\int_{a}^{b}} f = \overline{\int_{a}^{c}} f + \overline{\int_{c}^{b}} f$$

(and if f is integrable it therefore follows that

$$\int_{a}^{b} f = \int_{a}^{c} f + \int_{c}^{b} f)$$

Proof   We deal with the lower integrals first and, to start with, we'll assume that $a \leqslant b$ and that $c \in [a, b]$. Let L_{ab}, L_{ac} and L_{cb} be the set of under-estimates of integrals of f obtained by taking partitions of $[a, b]$, $[a, c]$ and $[c, b]$ respectively.

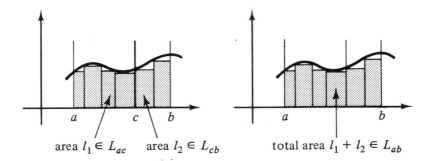

area $l_1 \in L_{ac}$ area $l_2 \in L_{cb}$ total area $l_1 + l_2 \in L_{ab}$

Isn't it clear without going into all the details that if $l_1 \in L_{ac}$ and $l_2 \in L_{cb}$ then $l_1 + l_2 \in L_{ab}$ as illustrated above? The interested reader can write out the formal details for him or herself. Hence

$l_1 + l_2 \in L_{ab}$	for each $l_1 \in L_{ac}$ and $l_2 \in L_{cb}$
$\therefore\ l_1 + l_2 \leqslant \sup L_{ab}$	for each $l_1 \in L_{ac}$ and $l_2 \in L_{cb}$
$\therefore\ \sup L_{ac} + \sup L_{cb} \leqslant \sup L_{ab}$	(this plausible step was in fact established in exercise 2 on page 13)

i.e.

$$\underline{\int_a^c} f + \underline{\int_c^b} f \leqslant \underline{\int_a^b} f$$

To prove the opposite inequality let l be any member of L_{ab} arising from a partition P of $[a, b]$ as illustrated on the left below. This partition may not actually use the number c. So let P' be the partition of $[a, b]$ consisting of P together with the number c and let l' be the under-estimate of the integral corresponding to P' (as shown in the middle figure below). Then, as we saw in the proof of the previous theorem, $l \leqslant l'$. Now (again leaving the full details to the interested reader) l' is clearly the sum of an $l_1 \in L_{ac}$ and an $l_2 \in L_{cb}$ as illustrated.

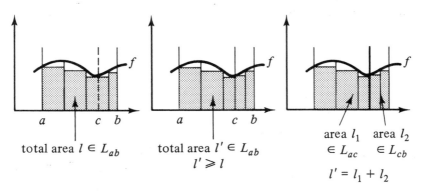

total area $l \in L_{ab}$ total area $l' \in L_{ab}$ area l_1 area l_2
                                    $l' \geqslant l$ $\in L_{ac}$ $\in L_{cb}$

$l' = l_1 + l_2$

Hence

$$l \leqslant l' = l_1 + l_2 \leqslant \sup L_{ac} + \sup L_{cb}$$
$$\underbrace{\qquad}_{\in L_{ac}} \quad \underbrace{\qquad}_{\in L_{cb}}$$

and so

$$l \leqslant \sup L_{ac} + \sup L_{cb} \quad \text{for each } l \in L_{ab}$$

and it follows that

$$\sup L_{ab} \leqslant \sup L_{ac} + \sup L_{cb}$$

i.e.

$$\underline{\int_a^b} f \leqslant \underline{\int_a^c} f + \underline{\int_c^b} f$$

We have now proved the inequality both ways round and so the theorem follows in this case.

What if the three numbers a, b and c are not in the order $a \leqslant c \leqslant b$? In fact the result follows easily in all cases. For example if $c < a \leqslant b$ then by what we have just proved (and by the rule for reversing the endpoints of the integration)

$$\underline{\int_c^b} f = \underline{\int_c^a} f + \underline{\int_a^b} f = -\underline{\int_a^c} f + \underline{\int_a^b} f$$

and the required result again follows.

But now what about the upper integrals? Either we can repeat the above proof making the necessary changes from 'lower' to 'upper' and from 'under-' to 'over-' etc, or we can note that it follows from the solution of exercise 5 above that

$$-\overline{\int_a^b} f = \underline{\int_a^b} (-f)$$

and so by the first part of the theorem applied to $-f$ we get

$$\overline{\int_a^b} f = -\underline{\int_a^b} (-f) = -\underline{\int_a^c} (-f) - \underline{\int_c^b} (-f) = \overline{\int_a^c} f + \overline{\int_c^b} f$$

Hence the result for upper integrals follows trivially.   □

Example   Let g be the function given by

$$g(x) = x|x| \quad x \in \mathbb{R}$$

Show that g is integrable over any interval $[a, b]$ and find

$$\int_{-1}^2 g$$

Solution   Let f be the integrable function given by $f(x) = x^2$ ($x \in \mathbb{R}$). Then

$$g(x) = \begin{cases} f(x) & x \geq 0 \\ -f(x) & x \leq 0 \end{cases}$$

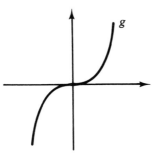

Now by the previous theorem (and from the value of f's integrals calculated in the previous exercises)

$$\int_{-1}^{2} g = \int_{-1}^{0} g + \int_{0}^{2} g = \int_{-1}^{0} (-f) + \int_{0}^{2} f$$

$$= \int_{-1}^{0} (-f) + \int_{0}^{2} f = -\int_{-1}^{0} f + \int_{0}^{2} f$$

$$= \tfrac{1}{3} \times (-1)^3 + \tfrac{1}{3} \times 2^3 = \tfrac{7}{3}$$

A similar result holds for g's upper integral and so g is integrable over $[-1, 2]$ with integral $\tfrac{7}{3}$. (The fact that g is integrable over any $[a, b]$ can be proved similarly and is also an immediate consequence of exercise 3 above since g is an increasing function.)   □

That example begins to show how any function which consists of pieces of integrable functions is itself integrable, and we begin to see that there is a very large collection of integrable functions. The next theorem is a giant leap because it shows that all continuous functions are integrable. Remember that integrals have been defined in terms of area – there has been no mention of gradients or of differentiation. It is therefore surprising perhaps that in the proof of the theorem we are going to use derivatives.

Theorem If f is continuous on $[a, b]$ then it is integrable over $[a, b]$.

Proof Since f is continuous on $[a, b]$ it is certainly bounded there (by the theorem on page 140). Hence for each $x \in [a, b]$ f is bounded on $[a, x]$ and the following lower and upper integrals are well defined:

$$\underline{\int_a^x} f \qquad \overline{\int_a^x} f$$

Now define functions F and G with domain $[a, b]$ by

$$F(x) = \underline{\int_a^x} f \quad \text{and} \quad G(x) = \overline{\int_a^x} f$$

(I know that from your previous experience with integrals you'd probably rather include a variable under the integral sign, for example preferring

$$\int_0^{\pi/2} \sin x \, dx \quad \text{to} \quad \int_0^{\pi/2} \sin$$

If you do prefer that well-established notation then in this case, as we've already used x as one of the endpoints of the integral, there will have to be a new variable under the integral sign: e.g.

$$F(x) = \underline{\int_a^x} f(t) \, dt \quad \text{and} \quad G(x) = \overline{\int_a^x} f(t) \, dt$$

Of course, as we commented before, the actual label chosen is irrelevant – the numerical value of $F(x)$ is unaffected.)

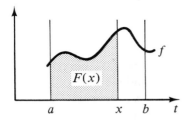

Now for x, $x_0 \in [a, b]$ what is $F(x) - F(x_0)$? For $x > x_0$ the following figure seems to imply that

$$F(x) - F(x_0) = (x - x_0) \times f(\gamma)$$

for some γ between x_0 and x: and that is exactly what we shall prove next.

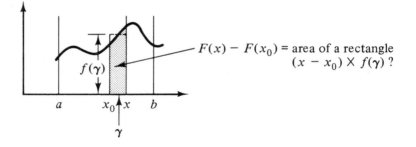

$F(x) - F(x_0)$ = area of a rectangle $(x - x_0) \times f(\gamma)$?

For any x and x_0 in $[a, b]$ we have by the previous theorem

$$F(x) - F(x_0) = \underline{\int_a^x f} - \underline{\int_a^{x_0} f} = \underline{\int_{x_0}^x f}$$

and

$$G(x) - G(x_0) = \overline{\int_a^x f} - \overline{\int_a^{x_0} f} = \overline{\int_{x_0}^x f}$$

Now by the comments on page 202, for $x > x_0$ we have

$$(x - x_0) \inf\{f(t): x_0 \leqslant t \leqslant x\} \leqslant \underline{\int_{x_0}^x f} \leqslant \overline{\int_{x_0}^x f}$$

$$\leqslant (x - x_0) \sup\{f(t): x_0 \leqslant t \leqslant x\}$$

and so as f is continuous

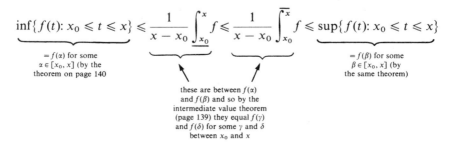

$$\underbrace{\inf\{f(t): x_0 \leqslant t \leqslant x\}}_{\substack{= f(\alpha) \text{ for some} \\ \alpha \in [x_0, x] \text{ (by the} \\ \text{theorem on page 140)}}} \leqslant \underbrace{\frac{1}{x - x_0} \int_{x_0}^x f}_{} \leqslant \underbrace{\frac{1}{x - x_0} \overline{\int_{x_0}^x f}}_{} \leqslant \underbrace{\sup\{f(t): x_0 \leqslant t \leqslant x\}}_{\substack{= f(\beta) \text{ for some} \\ \beta \in [x_0, x] \text{ (by} \\ \text{the same theorem)}}}$$

these are between $f(\alpha)$ and $f(\beta)$ and so by the intermediate value theorem (page 139) they equal $f(\gamma)$ and $f(\delta)$ for some γ and δ between x_0 and x

Similarly if $x < x_0$ we have (by the above with x and x_0 interchanged)

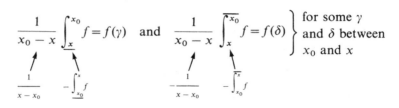

$$\frac{1}{x_0 - x} \underline{\int_x^{x_0} f} = f(\gamma) \quad \text{and} \quad \frac{1}{x_0 - x} \overline{\int_x^{x_0} f} = f(\delta) \left.\right\} \begin{array}{l} \text{for some } \gamma \\ \text{and } \delta \text{ between} \\ x_0 \text{ and } x \end{array}$$

So for **any** $x \neq x_0$ in $[a, b]$ we have

$$\frac{1}{x - x_0} \underline{\int_{x_0}^x} f = f(\gamma) \quad \text{and} \quad \frac{1}{x - x_0} \overline{\int_{x_0}^x} f = f(\delta) \left.\right\} \begin{array}{l} \text{for some } \gamma \\ \text{and } \delta \text{ between} \\ x_0 \text{ and } x \end{array}$$

and so

$$\frac{F(x) - F(x_0)}{x - x_0} = \frac{1}{x - x_0} \underline{\int_{x_0}^x} f = f(\gamma)$$

and

$$\frac{G(x) - G(x_0)}{x - x_0} = \frac{1}{x - x_0} \overline{\int_{x_0}^x} f = f(\delta)$$

for some γ and δ between x_0 and x. Therefore as $x \to x_0$ we have $\gamma, \delta \to x_0$ too and by the continuity of f we have

$$\underbrace{\lim_{x \to x_0} \frac{F(x) - F(x_0)}{x - x_0}}_{= F'(x_0)} = f(x_0) \quad \text{and} \quad \underbrace{\lim_{x \to x_0} \frac{G(x) - G(x_0)}{x - x_0}}_{= G'(x_0)} = f(x_0)$$

In other words F and G are differentiable functions with the same derivative throughout $[a, b]$ (namely $F'(x_0) = G'(x_0) = f(x_0)$ for each x_0). Therefore by exercise 7 on page 168 F and G differ by a constant; i.e. there exists a constant k with

$$F(x) = G(x) + k \quad \text{for each } x \in [a, b]$$

But

$$F(a) = \underline{\int_a^a} f = 0 \quad \text{and} \quad G(a) = \overline{\int_a^a} f = 0$$

and so the constant k is $F(a) - G(a)$ which is 0. Therefore $F(x) = G(x)$ for each $x \in [a, b]$ and in particular $F(b) = G(b)$; i.e.

$$\underline{\int_a^b} f = \overline{\int_a^b} f$$

and so (at last) we've seen that f is integrable over $[a, b]$.　□

That was quite a neat use of some of our earlier theorems on continuity, but wasn't it rather hard work without much real benefit (beyond learning that continuous functions are integrable, but so what)? As you may have realised it also brings us to the brink of a great result. Sometimes in mathematics the pieces fit together in such a beautiful way that it is nothing short of miraculous. For example we saw in chapter 1 that the exponential function and the logarithmic function, defined by apparently independent methods, turned out to be inverses of each other. No one with a spark of analysis in them can fail to marvel at that result the first time they see it. Unfortunately, however, it is hard to surprise you with such delights because you've already used the inter-relations between the topics at school (without actually having established them). So now wipe your minds clear, if you can, of your senior school mathematics. Imagine that in this book you have met differentiation and integration for the first time – one concerned with gradients and the other with areas. There seems no reason at all why the two topics should be linked. And yet stupendously it turns out that differentiation and integration are, in some sense, opposites of each other.

Example Let $f(x) = x^2$ $(x \in \mathbb{R})$ and define a function F by

$$F(x) = \int_1^x f \qquad x \in \mathbb{R}$$

Find F specifically in terms of x and show that it is differentiable with derivative equal to f.

Solution We've already seen in the previous exercises that

$$F(x) = \int_1^x f = \tfrac{1}{3}(x^3 - 1^3) = \tfrac{1}{3}(x^3 - 1) \quad x \in \mathbb{R}$$

Hence, by our standard results on differentiation, F is differentiable with derivative given by

$$F'(x) = x^2 = f(x) \qquad \qquad \Box$$

That example (and the proof of the previous theorem) show how the idea of an integral/area can be made into a function. Our final result (called the 'fundamental theorem of calculus') now confirms that integration is the 'opposite' of differentiation.

Corollary (*The fundamental theorem of calculus*) Let f be a continuous function whose domain D is an interval and let $a \in D$. Define a function F with domain D by

$$F(x) = \int_a^x f \qquad x \in D$$

Then F is differentiable and $F' = f$.

Proof This hardly needs a proof because we virtually showed this in the proof of the previous theorem. We saw there that if f is continuous on $[a, b]$ then f is integrable over $[a, b]$. Applying that to $[a, x]$ (or $[x, a]$) for any $x \in D$ we see that the F defined in the previous theorem becomes

$$F(x) = \int_a^x f = \int_a^x f$$

which is the same as the F in this corollary. Hence, as in the previous proof,

$$F'(x_0) = f(x_0) \quad \text{for each } x_0 \in D$$

and $F' = f$ as required.   □

Let us pause to digest that result. An integral

$$\int_a^b f$$

is some measure of an area (at school you probably called it a *definite integral*). If we make the upper endpoint of the integral into a variable x then we get a function F given by

$$F(x) = \int_a^x f$$

If f is continuous (as all the standard elementary functions are) then F is differentiable with derivative equal to f. Any function whose derivative is f is called an *indefinite integral* (or *primitive*) of f. Since any two functions with the same derivative differ by a constant it follows that any two indefinite integrals of f differ by a constant – so if you find one and write '$+k$' after it you have found them all. In particular any indefinite integral of f differs from the F defined above by a constant. It is this fact which you have taken for granted in the past to enable you to use your knowledge of differentiation to calculate some integrals.

Example   Calculate

$$\int_2^3 \frac{1}{x} \, dx$$

Solution   We saw in the chapter on differentiation that the function given by $F_1(x) = \log x$ differentiates to give $F_1'(x) = 1/x$: i.e. F_1 is an indefinite integral of $f(x) = 1/x$. You would probably already have jumped to the stage

$$\int_2^3 \frac{1}{x} \, dx = [\log x]_2^3 = \log 3 - \log 2 = \log \tfrac{3}{2}$$

But how is that justified? Since

$$F_1(x) = \log x \quad \text{and} \quad F_2(x) = \int_2^x f$$

are both indefinite integrals of f (which means that they have the same derivative, namely f) it follows that F_1 and F_2 differ by a constant; i.e.

$$F_2(x) = F_1(x) + k \quad \text{for each } x > 0$$

or

$$\int_2^x f = \log x + k \quad \text{for each } x > 0$$

In particular putting $x = 2$ gives $F_2(2) = 0$, $F_1(2) = \log 2$ and so $k = -\log 2$. Therefore

$$\int_2^x f = \log x - \log 2 \quad \text{for each } x > 0$$

and

$$\int_2^3 f = \int_2^3 \frac{1}{x}\, dx = \log 3 - \log 2 = \log \tfrac{3}{2}$$

as expected.

You will be able to show for yourselves in a similar way that in general if F differentiates to give f (i.e. F is an indefinite integral of f) then

$$\int_a^b f = F(b) - F(a) \qquad \qquad \Box$$

Seeing that integration is in some sense the opposite of differentiation also shows how the 'dx' notation again works well. For example if you are given that $y = f(x)$ satisfies

$$f'(x) = \frac{dy}{dx} = \frac{1}{x}$$

then as in the above example you deduce that (to within a constant)

$$y = f(x) = \log x$$

It seems that

$$\frac{dy}{dx} = \frac{1}{x} \implies \boxed{dy = \frac{1}{x}\, dx} \implies y = \int dy = \int \frac{1}{x}\, dx = \log x$$

But don't let anybody see that shaded bit – as we've said before it makes no sense to treat dy and dx as separate entities.

The fundamental theorem is a marvellous way to finish an introductory course in analysis. There are many directions in which these studies now lead, and indeed many loose ends still waiting to be tied up. So below you will find some more extensive exercises than usual (with just outline solutions at the back of the book). For the reader who wishes to study analysis further there are many excellent books. I must mention two in particular which have given me much pleasure over the years: K. G. Binmore's very readable *Mathematical analysis* (C.U.P.) covers the material in this book and takes it rather further and G. F. Simmons' *Introduction to topology and modern analysis* (McGraw-Hill) shows some of the delightful treats in store for those who pursue this subject further. And your college library shelves offer plenty of choice of books on analysis: this has, after all, simply been *yet another*.

Exercises

1   (Integration by parts.) For differentiable functions f and g with domain $[a, b]$ let the function F be given by

$$F(x) = f(x)g(x) - \int_a^x f'g \quad x \in [a, b]$$

Show that F is an indefinite integral of fg' and deduce that

$$\int_a^b fg' = [f(x)g(x)]_a^b - \int_a^b f'g$$

Hence evaluate

(i) $\displaystyle\int_0^{\pi/2} x \sin x \, dx$ (ii) $\displaystyle\int_1^e \log x \, dx$ (iii) $\displaystyle\int_0^{\pi/2} e^x \sin x \, dx$

2   (Integration by substitution.) Let g be a differentiable function with domain $[a, b]$ and which has a continuous derivative g'. Let f be a continuous function such that the composite function $f \circ g$ is well defined for each $x \in [a, b]$. Show that

$$\int_a^b (f \circ g) \times g' = \int_{g(a)}^{g(b)} f$$

(Using the fuller 'dx' notation and putting $g(x) = u$ would give

$$\int_a^b f(g(x)) \times g'(x) \, dx = \int_{(x=)a}^b f(g(x)) \frac{du}{dx} \, dx$$

$$= \int_{(u=)g(a)}^{g(b)} f(u) \, du$$

which once again looks very algebraically 'natural'.)

Use this method to find indefinite integrals of

(1) $2x \sin x^2$ (ii) $\dfrac{e^x}{1 + e^{2x}}$ (iii) $\tan x \left(x \in \left(-\dfrac{\pi}{2}, \dfrac{\pi}{2} \right) \right)$

3   (Improper integrals.) Our definition of an integral was for a bounded function over a closed bounded interval. We can extend this idea to other intervals and unbounded functions in the following sorts of ways, giving rise to the so-called 'improper integrals'

E.g. $\int_a^b f$ for an f bounded on $[a, x]$ for each $x \in [a, b)$ but not on $[a, b]$: let

$$F(x) = \int_a^x f \qquad x \in [a, b)$$

Then we say that $\int_a^b f$ exists if and only if $\lim_{x \to b} F(x)$ exists, and in that case the value of the integral is defined to be that limit.

e.g. $\int_a^\infty f$ where f is bounded on $[a, x]$ for each $x > a$: let

$$F(x) = \int_a^x f \qquad x > a$$

Then we say that $\int_a^\infty f$ exists if and only if $\lim_{x \to \infty} F(x)$ exists and in that case the value of the integral is defined to be that limit.

Decide whether the following improper integrals exist and evaluate those which do:

(i) $\displaystyle\int_0^1 \dfrac{1}{\sqrt{(1 - x^2)}} \, dx$ (ii) $\displaystyle\int_0^\infty \dfrac{x}{(1 + x^2)^2} \, dx$ (iii) $\displaystyle\int_0^1 \dfrac{1}{x} \, dx$

4   (The integral test for the convergence of series.) Let f be a decreasing function with domain $[1, \infty)$ and with $f(x) \geqslant 0$ for each $x \geqslant 1$. For each positive integer n let

$$x_n = f(1) + f(2) + f(3) + \cdots + f(n - 1) - \int_1^n f$$

Follow the techniques employed on pages 47–8 to show that the sequence (x_n) is increasing and bounded above and hence convergent. Deduce that the series

$$f(1) + f(2) + f(3) + \cdots$$

converges if and only if the integral

$$\int_1^\infty f$$

exists.

Use that result (or a slightly modified form of it) to test the convergence of the following series:

(i) $\displaystyle\sum_{n=1}^{\infty} \frac{1}{\sqrt{n}}$ (ii) $\displaystyle\sum_{n=2}^{\infty} \frac{1}{n \log n}$

(iii) $\displaystyle\sum_{n=3}^{\infty} \frac{1}{n \log n \log(\log n)}$ (iv) $\displaystyle\sum_{n=2}^{\infty} \frac{1}{n(\log n)^2}$

5   (An alternative form of the error term in a Taylor series.) Let f be differentiable more than n times on some interval containing 0 and x. Prove by induction on n that

$$f(x) = f(0) + xf'(0) + \frac{x^2}{2!}f''(0) + \cdots + \frac{x^n}{n!}f^{(n)}(0)$$

$$+ \frac{1}{n!}\int_0^x (x - t)^n f^{(n+1)}(t)\, dt$$

Hence show that for each $x \in (-1, 1]$ the difference between

$$\log(1 + x) \quad \text{and} \quad x - \frac{x^2}{2} + \frac{x^3}{3} - \frac{x^4}{4} + \cdots \pm \frac{x^n}{n}$$

converges to 0 as $n \to \infty$ (i.e. the Taylor series about $x = 0$ of $\log(1 + x)$ converges to $\log(1 + x)$ for $x \in (-1, 1]$).

6   (Integration of power series.) Show that if the power series

$$a_0 + a_1 x + a_2 x^2 + a_3 x^3 + \cdots$$

converges to $f(x)$ for $x \in (-r, r)$ then the 'integrated' series

$$a_0 x + \tfrac{1}{2}a_1 x^2 + \tfrac{1}{3}a_2 x^3 + \tfrac{1}{4}a_3 x^4 + \cdots$$

converges for each $x \in (-r, r)$ and its sum is $F(x)$ where F is an indefinite integral of f.

Hence show that

$$\tan^{-1} x = x - \frac{x^3}{3} + \frac{x^5}{5} - \frac{x^7}{7} + \cdots \qquad x \in (-1, 1)$$

and deduce that

$$\pi = 4(1 - \tfrac{1}{3} + \tfrac{1}{5} - \tfrac{1}{7} + \cdots)$$

7   (The irrationality of π.) Assume that π is the rational M/N where M and N are positive integers. Let k be a positive integer with $N^k \pi^{2k+1}/k!$ less than 1 (how do we know that such a k exists?) and let f be the function given by

$$f(x) = \frac{N^k}{k!} x^k (x - \pi)^k \qquad x \in [0, \pi]$$

(i) Show that f and all its derivatives have integer values at $x = 0$ and at $x = \pi$.

(ii) Prove that

$$0 < \int_0^\pi f(x) \sin x \, dx < 1$$

(iii) By repeatedly integrating by parts show that

$$\int_0^\pi f(x) \sin x \, dx$$

is an integer.

Deduce that π is irrational.

(Solutions on page 281)

Chapter 1

Page 5

1   Do you remember how to add fractions? For example

$$\frac{5}{6} + \frac{3}{8} = \frac{(5 \times 8) + (6 \times 3)}{48} = \frac{58}{48}$$

although obviously 48 is not the most economical denominator in this case. In general

$$\frac{m}{n} \pm \frac{p}{q} = \frac{mq \pm np}{nq} \qquad \frac{m}{n} \times \frac{p}{q} = \frac{mp}{nq} \qquad \frac{m}{n} \div \frac{p}{q} = \frac{mq}{np}$$

and in all cases the numerators and denominators of the answers are still integers. Hence all the answers are rational numbers.

Now let r be a rational number and s an irrational number. Just suppose for a moment that their sum is rational, r' say. Then

$$r + s = r' \quad \text{and} \quad s = r' - r$$

But we have just seen that the difference of two rational numbers is again rational. So it is impossible for the irrational s to be the difference of two rationals. This contradiction shows that $r + s$ must be irrational.

Similarly just suppose that the product of the non-zero rational r and the irrational s is the rational r'. Then

$$rs = r' \quad \text{and} \quad s = r' \div r$$

But this expresses the irrational s as the quotient of two rationals, which as we have just seen is impossible. Hence the product rs cannot be rational and it must be irrational.

The other parts follow by similar arguments.

2   The answer $4\frac{1}{8} = 4.125\,000\ldots$ is obtained from $33.000\,000\,0\ldots$ divided by 8. Similarly $3\frac{1}{7} = 3.142\,857\,142\,857\ldots$ can be calculated by dividing $22.000\,000\,0\ldots$ by 7. Although the 'art' of long division may not mean much to

some of you it is a convenient way of illustrating these processes here (and I'm going to assume that they *do* follow from primary arithmetic).

As soon as one of the remainders after the decimal point is zero the decimal answer stops (or continues with a string of zeros) as shown on the left. On the other hand, if we don't get a remainder of zero then as soon as one of the non-zero remainders has repeated itself the whole pattern of remainders and answers will repeat itself thereafter, as shown on the right:

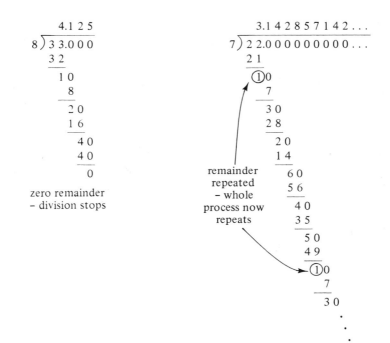

Since there are only $n - 1$ possible different non-zero remainders when dividing by n these repeated remainders must have occurred less than n places apart and the recurring string must be of less than n digits.

3   Since $1/74 = 0.0135135135\ldots$ we have

$$\tfrac{10}{74} = 0.135\,135\,135\ldots \quad \text{and} \quad \tfrac{10000}{74} = 135.135\,135\,135\ldots$$

The difference of these two then shows that $9990/74$ equals 135 and so 9990 is divisible by 74. In general, given the integer $n > 1$ we have, for some strings of digits A and B,

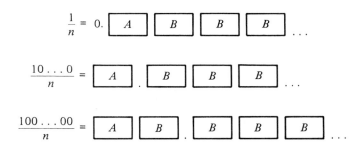

$$\frac{1}{n} = 0.\boxed{A}\ \boxed{B}\ \boxed{B}\ \boxed{B}\ \dots$$

$$\frac{10\dots0}{n} = \boxed{A}.\boxed{B}\ \boxed{B}\ \boxed{B}\ \dots$$

$$\frac{100\dots00}{n} = \boxed{A}\ \boxed{B}.\boxed{B}\ \boxed{B}\ \boxed{B}\ \dots$$

The difference of these last two shows that $99\dots9900\dots00/n$ equals an integer and therefore some multiple of n equals $99\dots9900\dots00$ as required.

Page 8

1   Assume that $\sqrt{3} = m/n$ where n are integers and the fraction m/n is assumed to be written in its sensible 'cancelled-down' form. Then

$$\frac{m^2}{n^2} = \left(\frac{m}{n}\right)^2 = 3 \quad \text{and} \quad m^2 = 3n^2$$

But then m is divisible by 3 and, by an argument very similar to that in the $\sqrt{2}$ case, you can deduce that n^2 is divisible by 3 and hence so is n. But then both m and n are divisible by 3 which contradicts the fact that the fraction m/n was in its sensible form. This contradiction shows that $\sqrt{3}$ cannot be rational.

     Similarly if $\sqrt[3]{2} = m/n$ then $m^3 = 2n^3$ and you can deduce that both m and n are divisible by 2.

2   In $m^2 = 2n^2$ the left-hand side is the product of $2M$ primes and the right-hand side (being $2 \times n \times n$) is the product of $2N + 1$ primes. But the number on either side of that equation has a unique representation as a product of primes: there will either be an odd number of them or an even number of them, but not both! So it is impossible to have integers m and n with $m^2 = 2n^2$ and so $\sqrt{2}$ cannot be rational.

     Now if $\sqrt{q}$ is the rational m/n then $qn^2 = m^2$. If you write this number $(=m \times m)$ as a product of primes then clearly each prime will occur an even number of times. It follows that when qn^2 is written as a product of primes each prime occurs an even number of times. But in that product the n^2 contributes primes in pairs. Hence when q itself is written as a product of primes each prime will occur an even number of times. Thus q is a perfect square and $\sqrt{q}$ is an integer.

3   (i) We are given that $\alpha > 0$, from which it is clear that $\beta > 0$, and we are given that $\alpha^2 > 2$ from which it follows that $\alpha/2$ is greater than $1/\alpha$. Hence

$$\alpha - \beta = \alpha - \left(\frac{\alpha}{2} + \frac{1}{\alpha}\right) = \frac{\alpha}{2} - \frac{1}{\alpha} > 0 \quad \text{and} \quad 0 < \beta < \alpha$$

Furthermore

$$\beta^2 - 2 = \left(\frac{\alpha}{2} + \frac{1}{\alpha}\right)^2 - 2 = \left(\frac{\alpha^2}{4} + 1 + \frac{1}{\alpha^2}\right) - 2 = \left(\frac{\alpha}{2} - \frac{1}{\alpha}\right)^2 > 0$$

and so $\beta^2 > 2$. Hence no matter which positive α you choose with $\alpha^2 > 2$ there is always a smaller such number.

(ii) Now if $\alpha > 0$ and $\alpha^2 < 2$ then $2/\alpha > 0$ and $(2/\alpha)^2 > 2$. The process is illustrated below with the line representing the real numbers and the heavy shading representing those numbers whose square is less than 2.

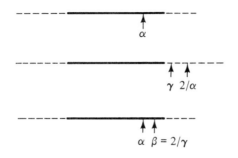

We can now apply (i) to $2/\alpha$ to deduce the existence of a positive number γ with $\gamma < 2/\alpha$ and $\gamma^2 > 2$. Let $\beta = 2/\gamma$. Then

$$\beta = \frac{2}{\gamma} > \alpha$$

and

$$\beta^2 = \left(\frac{2}{\gamma}\right)^2 = \frac{4}{\gamma^2} < 2$$

So there is no biggest positive number (and hence no biggest number) whose square is less than 2.

Page 13

1   For any number u the integer $[u]$ is chosen to satisfy $0 \leqslant u - [u] < 1$. Hence if u is a positive number then $[u] + 1$ is a positive integer larger than u. It follows that if you give me any number u then I can find a member of $\mathbb{N}$ which is larger than u. Hence no number u is an upper bound of $\mathbb{N}$. Therefore $\mathbb{N}$ is not bounded above.

2   (i) You will probably have guessed that the set A of the prime numbers is not bounded above. The formal verification uses the fact (known almost from primary school!) that given any positive integer n there exists a prime number bigger than n (for example any prime factor of $n! + 1$). Hence given any member of $\mathbb{N}$ there is a bigger member of A. The fact that A is not bounded above therefore follows from the fact that $\mathbb{N}$ is not bounded above.

The other sets are bounded above and their suprema are given by

$$\sup B = \tfrac{1}{2} \qquad \sup C = 1 \qquad \sup D = 1$$

(ii) There are several ways of proving this result but here's quite a neat one:

$a + b \leqslant \alpha$ for each $a \in A$ and $b \in B$

$\therefore \ a \leqslant \underbrace{\alpha - b}$ for each $a \in A$ (and fixed $b \in B$)

$\therefore$ an upper bound for A – but sup A is the least such

$\therefore \ \sup A \leqslant \alpha - b$ for each $b \in B$

$b \leqslant \underbrace{\alpha - \sup A}$ for each $b \in B$

$\therefore$ an upper bound for B – but sup B is the least such

$\therefore \ \sup B \leqslant \alpha - \sup A$

and

$\sup A + \sup B \leqslant \alpha$

as required.

3   (i) A set M is *bounded below* if there exists a number l (called a *lower bound* of M) with $l \leqslant m$ for each $m \in M$. In this case let L be the (non-empty) set of lower bounds of M and by the completeness axiom let α be 'between' L and M. Then it is easy to check that α is a lower bound of M and is the biggest of all such lower bounds. Hence α is the biggest lower bound (or infimum) of M.

(ii) In exercise 2(i) all four sets are bounded below and their infima are

$$\inf A = 2 \quad \inf B = 0 \quad \inf C = \tfrac{1}{2} \quad \inf D = \tfrac{1}{2}$$

(iii) In general if L and M are any non-empty sets with $l \leqslant m$ for each $l \in L$ and $m \in M$ then the numbers 'between' L and M are precisely those which are both upper bounds of L and lower bounds of M. So it follows that if α is the only number between the two sets then α must be the least upper bound of L. (For if $\beta < \alpha$ then β is not between L and M and yet β is still a lower bound of M, so it's clear that β is not an upper bound of L.) Similarly α must be the greatest lower bound of M.

(iv) Clearly if $\alpha \leqslant a$ for each $a \in A$ then $-\alpha \geqslant -a$ for each $a \in A$ and so $-\alpha \geqslant b$ for each $b \in B$. It follows that if α is a lower bound of A then $-\alpha$ is an upper bound of B: it is also easy to check that if β is not a lower bound of A then $-\beta$ is not an upper bound of B.

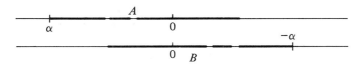

In addition if α is the *greatest* lower bound (or infimum) of A then we claim that $-\alpha$ is the *least* upper bound (or supremum) of B. For if $\beta < -\alpha$ then $-\beta > \alpha$, $-\beta$ is not a lower bound of A, and β is not an upper bound of B. Hence $-\alpha$ is indeed the lowest of all B's upper bounds as required.

4   (i) This is our first result using the phrase 'if and only if'. We have two statements

(1) there exists a number b with $-b \leqslant e \leqslant b$ for each $e \in E$;

(2) the set E is bounded.

We will show that if (2) is true then (1) is true; i.e. **(1) if (2)**. We will also show that if (1) is true then so is (2), or in other words (1) can only happen if (2) follows; i.e. **(1) only if (2)**.

Putting those two highlighted statements together we get the mathematician's stock phrase '(1) if and only if (2)' which simply means that each implies the other and the two things always occur together.

So to prove the result we first assume (2) and deduce (1). By that assumption the set E is bounded and so it is bounded above (by u say) *and* bounded below (by l say) and we have $l \leqslant e \leqslant u$ for each $e \in E$. Now let b be the larger of u and $-l$. Then for each $e \in E$

$$-b \leqslant l \leqslant e \leqslant u \leqslant b \quad \text{and} \quad -b \leqslant e \leqslant b$$

We have therefore established (1) as required.

Conversely we now assume (1) and deduce (2). By (1) there exists a number b with

$$-b \leqslant e \leqslant b \quad \text{for each } e \in E$$

Therefore E is bounded above by b, bounded below by $-b$, and hence bounded. So property (2) follows as required.

We have thus shown that '(1) if and only if (2)'. In future proofs of this type we shan't spell out the steps quite so pedantically.

(ii) Now if E and E' are bounded then there exist numbers b and b' such that

$$-b \leqslant e \leqslant b \text{ for each } e \in E \quad \text{and} \quad -b' \leqslant e' \leqslant b' \quad \text{for each } e' \in E'$$

We may assume that $b \geqslant b'$ (for if not an identical argument with E and E' reversed would work). Then if x is in E we have

$$-b \leqslant x \leqslant b$$

and if x is in E' we have

$$-b \leqslant -b' \leqslant x \leqslant b' \leqslant b$$

Hence whether x is in E or in E' it follows that

$$-b \leqslant x \leqslant b$$

and so the union $E \cup E'$ of the two bounded sets is also bounded.

5   Assume that $x < y$ (a similar argument works in the other case). Then

$$x = x + 0(y - x) < \underbrace{x + \beta(y - x)}_{= (1 - \beta)x + \beta y} < x + 1(y - x) = y$$

and (bearing in mind our results about adding and multiplying rationals and irrationals from exercise 1 on page 5):

(a) if x and y are rationals and $\beta = \frac{1}{2}$ then $(1 - \beta)x + \beta y$ is a rational number (midway) between x and y;

(b) if x and y are rationals and $\beta = 1/\sqrt{2}$ (which is irrational) then $x + \beta(y - x)$ is an irrational number between x and y;

(c) if one of x and y is rational and the other irrational and $\beta = \frac{1}{2}$ then $(1 - \beta)x + \beta y$ is an irrational number (again midway) between x and y.

(In fact for $x \neq y$ any number z may be written as

$$z = \frac{y - z}{y - x} x + \frac{z - x}{y - x} y = \alpha x + \beta y$$

where α and β add up to 1. In particular each number between x and y is of the form $z = \alpha x + \beta y$ for some positive α and β with $\alpha + \beta = 1$.)

6   (i) The number attending the match must be between 22 500 and 23 499 (inclusive), for by convention 23 500 would be rounded up to 24 000. Hence the largest possible number of people at the match is 23 499. That is the correct answer to a perfectly legitimate question.

(ii) An angle of $48.5°$ will be rounded up to $49°$ but angles of

$$48.49°, \; 48.499°, \; 48.4999°, \; 48.4999°, \ldots$$

will all be rounded down to $48°$. Hence there is no *largest* angle which equals $48°$ to the nearest degree.

Page 18

1   The empty set $\varnothing$ is a slight oddity in the sense that it is bounded above (by any number at all) but has no least upper bound and no maximum member. (Note that our theorem about sets which are bounded above having a least upper bound expressly restricted itself to non-empty sets.)

For some of the other sets you have to add some terms of a geometric progression (or simply spot the fairly obvious pattern of the answers). Only B and D have maxima, namely $\frac{1}{2}$ and 1 respectively. In the other cases

sup $A = \sqrt{2}$ which as we've seen is not in $\mathbb{Q}$ (and hence not in A)
sup $C = 2$ which is not a member of C

2   If a set E has maximum member m (which must be in the set) then clearly m is an upper bound of the set. No upper bound of the set can be less than m and so m is the supremum of the set.

Conversely, if sup E is in the set E then sup E is an upper bound of the set which is contained in the set; i.e. sup E is the maximum member of the set.

(Here we have seen two properties

(1) E has a supremum which is contained in E;
(2) E has a maximum member;

and shown that each implies the other, again showing that '(1) if and only if (2)'.)

A set has a minimum if and only if it has an infimum which is contained in the set. The proof is very similar to that above for the maximum (with 'maximum' replaced by 'minimum', 'upper' by 'lower' etc.). Just the sets A and C have minima, the number 1 in each case. In the other cases

inf $B = 0 \notin B$ and   inf $D = \frac{1}{2} \notin D$

3   The set $E = \{n \in \mathbb{Z}: n < y\}$ is non-empty (it contains x for example) and it is bounded above (by y for example). Hence by the theorem on page 14 the set E has a maximum member, M say. Assume for the moment that $M \leqslant x$. But then

$$M + 1 \leqslant x + 1 < y$$

and so the integer $M + 1$ is still in the set E. This is impossible because because M was the largest such member, and this contradiction shows that our assumption that $M \leqslant x$ was wrong. Hence $M > x$ and M is an integer between x and y.

4        $\mathbb{N}$ is not bounded above and so there exists an integer N with $N > 1/(y - x)$. Hence $Ny > Nx + 1$ and by the previous exercise applied to the numbers Nx and Ny there exists an integer M with $Nx < M < Ny$. Then the rational M/N lies between x and y.

In exercise 5 on page 14 we saw that between a rational and any other number there is an irrational. Hence for any $x < y$ we have

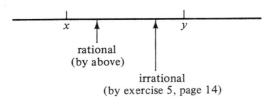

5   (i) We have to prove that

$$\frac{1}{1 \times 2} + \frac{1}{2 \times 3} + \frac{1}{3 \times 4} + \cdots + \frac{1}{n(n + 1)} = 1 - \frac{1}{n + 1}$$

for each positive integer n and to do this we prove the two steps outlined in the principle of induction on page 16:

(I) In the case $n = 1$ the result is trivial since both sides of the equation equal $\frac{1}{2}$.
(II) Assume that the result is true for some positive integer $k - 1$.

Then

$$\frac{1}{1 \times 2} + \frac{1}{2 \times 3} + \frac{1}{3 \times 4} + \cdots + \frac{1}{(k - 1)k} = 1 - \frac{1}{k}$$

Hence

$$\frac{1}{1 \times 2} + \frac{1}{2 \times 3} + \cdots + \frac{1}{k(k + 1)} = \left(\frac{1}{1 \times 2} + \frac{1}{2 \times 3} + \cdots + \frac{1}{(k - 1)k} \right) + \frac{1}{k(k + 1)}$$

$$= \left(1 - \frac{1}{k} \right) + \frac{1}{k(k + 1)} = 1 - \frac{(k + 1) - 1}{k(k + 1)}$$

$$= 1 - \frac{1}{k + 1}$$

and we have established the corresponding result for the case $n = k$. It follows by the principle of mathematical induction that the result is true for each positive integer n.

(ii) Again (I) is trivial. In (II) assuming the result for $k - 1$ and trying to deduce it for k gives

$$1^2 + 2^2 + \cdots + k^2 = (1^2 + 2^2 + \cdots + (k-1)^2) + k^2$$
$$= \tfrac{1}{6}(k-1)k(2(k-1)+1) + k^2$$
$$= \tfrac{1}{6}k((k-1)(2k-1) + 6k)$$
$$= \tfrac{1}{6}k(k+1)(2k+1)$$

as required.

6   (i) Again we follow the two steps necessary in the use of induction.

(I) In the case $n = 1$ the (in)equality is trivial, both sides being equal to $1 + x$.

(II) Now assume that the inequality holds for some positive integer $k - 1$. Then

$$(1 + x)^{k-1} \geqslant 1 + (k-1)x$$
$$\therefore \quad (1 + x)^k = (1 + x)^{k-1}(1 + x)$$
$$\ast \geqslant (1 + (k-1)x)(1 + x)$$
$$= 1 + x + (k-1)x + (k-1)x^2$$
$$= 1 + kx + (k-1)x^2 \geqslant 1 + kx$$

and we have derived the result in the case of $n = k$.

Bernoulli's inequality has therefore been established by the principle of mathematical induction.

    (Where did we use the fact that $x \geqslant -1$? To get $\ast$ we multiplied both sides of a known inequality by $1 + x$: that is only valid if $1 + x \geqslant 0$.)

    (ii) Note firstly that the 'binomial coefficients' satisfy

$$\binom{n}{0} = 1 \qquad \binom{n}{n} = 1$$

and

$$\binom{k-1}{r-1} + \binom{k-1}{r} = \frac{(k-1)!}{(r-1)!(k-r)!} + \frac{(k-1)!}{r!(n-1-r)!}$$
$$= \frac{(k-1)!(r+k-r)}{r!(k-r)!} = \binom{k}{r}$$

(which forms the basis of 'Pascal's triangle' of the binomial coefficients:

$$1$$
$$1 \quad 1$$
$$1 \quad 2 \quad 1$$
$$1 \quad 3 \quad 3 \quad 1$$
$$\vdots \qquad\qquad)$$

We now establish the two steps necessary in a proof by induction:

(I) In the case $n = 1$ the left-hand side of the required result is $(1 + x)$ and the right-hand side is $\binom{1}{0} + \binom{1}{1}x$, which is the same thing.

(II) Now we assume that the result holds for some positive integer $k - 1$ and try to deduce it for k:

$$(1 + x)^{k-1} = \binom{k-1}{0} + \binom{k-1}{1}x + \cdots + \binom{k-1}{r-1}x^{r-1}$$
$$+ \binom{k-1}{r}x^r + \cdots + \binom{k-1}{k-1}x^{k-1}$$

$\therefore \quad (1 + x)^k = (1 + x)^{k-1} \times (1 + x)$

$$= \left(\binom{k-1}{0} + \binom{k-1}{1}x + \cdots + \binom{k-1}{r-1}x^{r-1} \right.$$
$$\left. + \binom{k-1}{r}x^r + \cdots + \binom{k-1}{k-1}x^{k-1} \right) \times (1 + x)$$

$$= \binom{k-1}{0} + \left(\binom{k-1}{0} + \binom{k-1}{1} \right)x + \cdots$$
$$+ \left(\binom{k-1}{r-1} + \binom{k-1}{r} \right)x^r + \cdots + \binom{k-1}{k-1}x^k$$

$$= \binom{k}{0} + \binom{k}{1}x + \cdots + \binom{k}{r}x^r + \cdots + \binom{k}{k}x^k$$

as required. Hence the result has been proved by induction.

7   (i) Again we establish steps (I) and (II):

(I) In the case $n = 1$ the (in)equality is trivial since both sides are equal to $1\frac{1}{2}$.

(II) Assume that the result holds for some positive integer $k - 1$. Then

$$1 + \frac{1}{2} + \frac{1}{3} + \frac{1}{4} + \cdots + \frac{1}{2^{k-2}} \geqslant \tfrac{1}{2}((k-1) + 1) = \tfrac{1}{2}k$$

and so trying to derive the result for $n = k$ we deduce that

$$1 + \frac{1}{2} + \frac{1}{3} + \cdots + \frac{1}{2^{k-1}} = \left(1 + \frac{1}{2} + \cdots + \frac{1}{2^{k-2}} \right)$$
$$+ \underbrace{\left(\frac{1}{2^{k-2} + 1} + \cdots + \frac{1}{2^{k-1}} \right)}_{2^{k-2} \text{ terms } \geqslant \frac{1}{2^{k-1}}} \geqslant \tfrac{1}{2}k + \tfrac{1}{2} = \tfrac{1}{2}(k+1)$$

which is exactly what we want.

Hence the inequality is established by induction.

(ii)

(I) In the case $n = 1$ both sides of the inequality are 1 and so the result holds for $n = 1$.

(II) Assume that the results holds for some positive integer $k - 1$. Then

$$1 + \frac{1}{2^r} + \frac{1}{3^r} + \frac{1}{4^r} + \cdots + \frac{1}{(2^{k-1} - 1)^r} \leqslant \frac{1 - (\tfrac{1}{2})^{(r-1)(k-1)}}{1 - (\tfrac{1}{2})^{r-1}}$$

and so trying to derive the result for $n = k$ we deduce that

$$1 + \frac{1}{2^r} + \frac{1}{3^r} + \cdots + \frac{1}{(2^k - 1)^r} = \left(1 + \frac{1}{2^r} + \frac{1}{3^r} + \cdots + \frac{1}{(2^{k-1} - 1)^r}\right)$$

$$+ \left(\underbrace{\frac{1}{(2^{k-1})^r} + \frac{1}{(2^{k-1} + 1)^r} + \cdots + \frac{1}{(2^k - 1)^r}}_{2^{k-1} \text{ terms each} \leqslant \frac{1}{(2^{k-1})^r}}\right)$$

$$\leqslant \frac{1 - (\frac{1}{2})^{(r-1)(k-1)}}{1 - (\frac{1}{2})^{r-1}} + (\tfrac{1}{2})^{(r-1)(k-1)}$$

$$= \frac{1 - (\frac{1}{2})^{(r-1)k}}{1 - (\frac{1}{2})^{r-1}}$$

which is exactly what we want.

Hence the inequality is established by induction.

Page 22

1   Remember that your calculator or computer only produces approximate answers. In this case my calculator gave the approximate values of both the 10th and 11th terms as 2.718 282. (A short program also gave this value for the 1000th term!) So that seems a reasonable approximation for e.

2   To see that P is bounded above note from exercise 5 on page 19 that for each $n > 1$

$$1 + \tfrac{1}{4} + \tfrac{1}{9} + \tfrac{1}{16} + \cdots + \frac{1}{n^2} \leqslant 1 + \frac{1}{1 \times 2}$$

$$+ \frac{1}{2 \times 3} + \frac{1}{3 \times 4} + \cdots + \frac{1}{(n-1)n} = 2 - \frac{1}{n} < 2$$

and hence that each member of P is less than $\sqrt{(6 \times 2)} = \sqrt{12}$. Therefore P is bounded above and has a supremum. The 500th and 1000th terms of P are approximately 3.139681 and 3.140638, suggesting that the supremum is approximately 3.141.

3   Here the nth member of the set E is

$$1 + \tfrac{1}{2} + \tfrac{1}{3} + \tfrac{1}{4} + \cdots + \frac{1}{n}$$

A programmable calculator gave the 1000th and 1001st members as approximately 7.48547 and 7.48646, and you might begin to think that the set will be bounded above (by the number 8 for example?). But the 2000th member is approximately 8.17836, the 4000th member is approximately 8.87139, and the 8000th member 9.56447. So although progress is slow it seems that the members of this set may go up and up without bound. In fact (from exercise 7(i) on page 19) for each integer n

$$1 + \tfrac{1}{2} + \tfrac{1}{3} + \tfrac{1}{4} + \cdots + \frac{1}{2^{n-1}} \geqslant \tfrac{1}{2}(1 + n)$$

Hence the 2nd member of the set is at least $1\frac{1}{2}$, the 4th member is at least 2, the 8th member is at least $2\frac{1}{2}$, the 16th at least 3, the 32nd at least $3\frac{1}{2}$, the 64th at least 4,... and it follows that the set E is, surprisingly perhaps, not bounded above.

However for $r > 1$ the number $r - 1$ is positive and we now show that the set

$$E' = \left\{ 1, \quad 1 + \frac{1}{2^r}, \quad 1 + \frac{1}{2^r} + \frac{1}{3^r}, \quad 1 + \frac{1}{2^r} + \frac{1}{3^r} + \frac{1}{4^r}, \cdots \right\}$$

is bounded above. From exercise 7(ii) on page 19 we see that the nth member of E' satisfies

$$1 + \frac{1}{2^r} + \frac{1}{3^r} + \cdots + \frac{1}{n^r} \leqslant 1 + \frac{1}{2^r} + \frac{1}{3^r} + \cdots + \frac{1}{(2^n - 1)^r}$$

$$\leqslant \frac{1 - (\frac{1}{2})^{(r-1)n}}{1 - (\frac{1}{2})^{r-1}} < \frac{1}{1 - (\frac{1}{2})^{r-1}}$$

and hence the set E' is bounded above (by $1/(1 - (\frac{1}{2})^{r-1})$) as required.

It is interesting to note that, although the set of numbers

$$1, \quad 1 + \tfrac{1}{2}, \quad 1 + \tfrac{1}{2} + \tfrac{1}{3}, \quad 1\!\!\!/ + \tfrac{1}{2} + \tfrac{1}{3} + \tfrac{1}{4}, \cdots$$

is not bounded above, the set of numbers

$$1, \quad 1 + \frac{1}{2^{1.001}}, \quad 1 + \frac{1}{2^{1.001}} + \frac{1}{3^{1.001}}, \quad 1 + \frac{1}{2^{1.001}} + \frac{1}{3^{1.001}} + \frac{1}{4^{1.001}}, \cdots$$

is bounded above (by 1444 for example, since $1/(1 - (\frac{1}{2})^{0.001}) \approx 1443.2$).

Chapter 2

Page 28

1   (i) The gradient of the chord from $P(2, \frac{1}{2})$ to $Q(3, \frac{1}{3})$ is given by

$$\frac{\text{change in } y}{\text{change in } x} = \frac{\frac{1}{3} - \frac{1}{2}}{3 - 2} = -\frac{1}{6}$$

Repeating the process for Q equal, in turn, to (2.5, 1/2.5), (2.2, 1/2.2), (2.1, 1/2.1), (2.01, 1/2.01), (2.001, 1/2.001), ... gives the sequence of approximate values

$$-0.166667, \ -0.02, \ -0.227273, \ -0.238095, \ -0.248756, \ -0.249875, \ldots$$

(Your A-level calculus techniques, as yet unjustified here, will show you that $g'(2) = -0.25$.)

(ii) The figure showed the rectangles obtained when the area is divided into 6 strips. Their total area is

$$\tfrac{1}{6}(\tfrac{6}{7} + \tfrac{6}{8} + \tfrac{6}{9} + \tfrac{6}{10} + \tfrac{6}{11} + \tfrac{6}{12})$$

which equals approximately 0.6532107. Repeating this procedure for 8, 10 and 12 strips etc. gives a sequence of answers

$$0.6532107, \ 0.6628719, \ 0.6687714, \ 0.6727475, \ldots$$

(Again your A-level calculus, as yet unjustified, would lead you to expect an answer of $\log_e 2$ or ln 2, whatever that might mean.)

2   When the area under the graph of $g(x) = 1/x$ is divided into 10 strips the area of the rectangles is given by

$$h\left(\frac{1}{1+h} + \frac{1}{1+2h} + \frac{1}{1+3h} + \cdots + \frac{1}{1+10h}\right)$$

where h is $(e-1)/10$. My calculator gave the value of this expression as approximately 0.9478. Use of a short program gave the value of the equivalent expression in the case of 1000 strips as approximately 0.9994572. Then repeatedly doubling the number of strips 2000, 4000 etc gave the following sequence converging to the required area:

0.9994572, 0.9997291, 0.9998639, 0.9999323, 0.9999663, ...

It would seem, therefore, that the area could well be 1. (School calculus would give the answer $\log_e e$, whatever that means.)

3   Taking $x_1 = 10$ and using the rule

$$x_n = \frac{x_{n-1}}{2} + \frac{1}{x_{n-1}}$$

gives values $x_2 = 5.1$, $x_3 = 2.746\,078, \ldots$, etc, and the sequence of approximate answers

10, 5.1, 2.746078, 1.737194, 1.444238, 1.414525, 1.414213, ...

apparently tending towards $\sqrt{2}$.

Page 36

1   Let the nth term of the given sequence be x_n. Then for n odd we have

$$x_n = 1 - \tfrac{1}{2} + \cdots - \frac{1}{2^{n-2}} + \frac{1}{2^{n-1}} > 1 - \tfrac{1}{2} + \cdots$$

$$\underbrace{- \frac{1}{2^{n-2}} + \frac{1}{2^{n-1}} - \frac{1}{2^n} + \frac{1}{2^{n+1}}}_{<0} = x_{n+2}$$

and so the sequence $x_1, x_3, x_5, \ldots$ is decreasing. Similarly, for n even we have

$$x_n = 1 - \tfrac{1}{2} + \cdots + \frac{1}{2^{n-2}} - \frac{1}{2^{n-1}} < 1 - \tfrac{1}{2} + \cdots$$

$$\underbrace{+ \frac{1}{2^{n-2}} - \frac{1}{2^{n-1}} + \frac{1}{2^n} - \frac{1}{2^{n+1}}}_{>0} = x_{n+2}$$

and so the sequence $x_2, x_4, x_6, \ldots$ is increasing.

It is possible to use similar inequalities to show that every even-numbered term is less than every odd-numbered term. Alternatively you can use the sum of a geometric progression to show that for m even and n odd

$$x_m = \tfrac{2}{3}\left(1 - \frac{1}{2^m}\right) < \tfrac{2}{3} < \tfrac{2}{3}\left(1 + \frac{1}{2^n}\right) = x_n$$

It also seems from those sums that as m and n increase the terms of the sequence approach $\frac{2}{3}$.

The graph of the sequence is

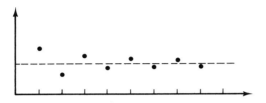

2   To see informally that the sequence $x_1, x_2, x_3, \ldots$ is decreasing with the terms shrinking towards 0 you could note that the triangles are getting slimmer and slimmer. More formally you could note that the sequence is

$$\tfrac{1}{2}\sin\frac{\pi}{2},\ \tfrac{1}{2}\sin\frac{\pi}{4},\ \tfrac{1}{2}\sin\frac{\pi}{8},\ \tfrac{1}{2}\sin\frac{\pi}{16},\ldots$$

and from your basic knowledge of the graph of the sine function note that these terms are decreasing and 'tending towards 0.'

To see that the sequence

$$x_1,\ 2x_2,\ 4x_3,\ 8x_4,\ldots$$

is increasing recall Archimedes' construction of π. The terms of this new sequence represent the areas of the figures illustrated and they would therefore seem to be increasing towards $\pi/4$.

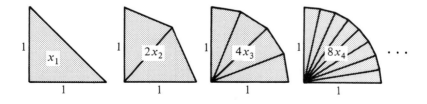

3   The equation of the line PQ is

$$y - f(x_0) = \frac{f(x_1) - f(x_0)}{x_1 - x_0}(x - x_0)$$

In particular when $x = x_2$ we have

$$y = \frac{f(x_1) - f(x_0)}{x_1 - x_0}(x_2 - x_0) + f(x_0) = \frac{f(x_0)(x_1 - x_2) + f(x_1)(x_2 - x_0)}{x_1 - x_0}$$

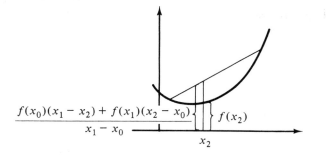

$$\frac{f(x_0)(x_1 - x_2) + f(x_1)(x_2 - x_0)}{x_1 - x_0} \quad \} f(x_2)$$

$$x_2$$

Now if f is convex then the chord PQ lies on or above the graph, and for $x_0 < x_2 < x_1$ we have

$$f(x_2) \leqslant \frac{f(x_0)(x_1 - x_2) + f(x_1)(x_2 - x_0)}{x_1 - x_0}$$

which rearranges to give

$$\text{gradient of } PQ' = \frac{f(x_2) - f(x_0)}{x_2 - x_0} \leqslant \frac{f(x_1) - f(x_0)}{x_1 - x_0} = \text{gradient of } PQ$$

Hence each time a new point Q' is chosen between P and Q it follows that the gradient of the chord stays the same or is reduced. Hence a sequence of gradients obtained in this way is decreasing.

4   If f is a concave function then (looking at it reflected in the x-axis) the function $-f$ (i.e. the function whose value at x is $-f(x)$) is convex. Hence if the sequence of gradients for f is $s_1, s_2, s_3, \ldots$ then it is easy to check that the sequence of gradients for the convex function $-f$ is $-s_1, -s_2, -s_3, \ldots$. By exercise 3 this sequence is decreasing and hence the sequence $s_1, s_2, s_3, \ldots$ is increasing.

5   Let the given sequence be $x_1, x_2, x_3, \ldots$. Then $x_n = (1 + 1/n)^n$. Now by Bernoulli's inequality we have

$$(1 + x)^n \geqslant 1 + nx \quad \text{(for } x \geqslant -1)$$

Putting $x = -1/n^2$ we get for $n > 1$

$$\left(1 + \frac{1}{n}\right)^n \left(1 - \frac{1}{n}\right)^n = \left(1 + \left(-\frac{1}{n^2}\right)\right)^n \geqslant 1 + n \times \left(-\frac{1}{n^2}\right) = 1 - \frac{1}{n}$$

$$\therefore \; x_n = \left(1 + \frac{1}{n}\right)^n \geqslant \frac{1 - 1/n}{(1 - 1/n)^n} = \frac{1}{(1 - 1/n)^{n-1}} = \frac{1}{((n-1)/n)^{n-1}}$$

$$= \left(\frac{n}{n-1}\right)^{n-1} = \left(1 + \frac{1}{n-1}\right)^{n-1} = x_{n-1}$$

and it follows that the sequence $x_1, x_2, x_3, \ldots$ is increasing.

My calculator gives the approximate value of 2.7182818 for both the 10000th and 20000th terms, implying perhaps that the terms of the sequence 'tend towards' e. In fact the terms of this increasing sequence never exceed e since

$$x_n = \left(1 + \frac{1}{n}\right)^n = 1 + n\frac{1}{n} + \frac{n(n-1)}{2!}\left(\frac{1}{n}\right)^2 + \frac{n(n-1)(n-2)}{3!}\left(\frac{1}{n}\right)^3 + \cdots$$

$$+ \frac{n(n-1)\ldots 1}{n!}\left(\frac{1}{n}\right)^n \leqslant 1 + 1 + \frac{1}{2!} + \frac{1}{3!} + \cdots + \frac{1}{n!} < e$$

Page 42

1   In the sequence $1, \frac{2}{3}, \frac{3}{5}, \frac{4}{7}, \frac{5}{9}, \frac{6}{11}, \ldots$ the nth term is given by

$$x_n = \frac{n}{2n-1} = \frac{1}{2} + \frac{1}{4n-2}$$

So the difference between x_n and the hoped-for limit of $\frac{1}{2}$ is $1/(4n-2)$. Given any $\varepsilon > 0$ that difference will be less than ε provided that $4n - 2$ is larger than $1/\varepsilon$. So let N be any integer larger than $\frac{1}{4}(1/\varepsilon + 2)$. Then $4N - 2 > 1/\varepsilon$ and

$$0 < \frac{1}{4N-2} < \varepsilon \quad \text{and} \quad 0 < \frac{1}{4(N+1)-2} < \varepsilon \quad \text{and}$$

$$0 < \frac{1}{4(N+2)-2} < \varepsilon \quad \text{and} \quad \ldots$$

It follows that

$$\tfrac{1}{2} - \varepsilon < x_N < \tfrac{1}{2} + \varepsilon \quad \text{and} \quad \tfrac{1}{2} - \varepsilon < x_{N+1} < \tfrac{1}{2} + \varepsilon \quad \text{and}$$

$$\tfrac{1}{2} - \varepsilon < x_{N+2} < \tfrac{1}{2} + \varepsilon \quad \text{and} \quad \ldots$$

(with the left-hand inequality in each case being rather trivially satisfied). Hence for any given $\varepsilon > 0$ we have found an N as required and it follows from the definition that the sequence converges to $\frac{1}{2}$.

2   Let $\varepsilon > 0$. To show that the sequence

$$\frac{1}{\log_{10} 2}, \quad \frac{1}{\log_{10} 3}, \quad \frac{1}{\log_{10} 4}, \quad \frac{1}{\log_{10} 5}, \ldots, \frac{1}{\log_{10}(n+1)}, \cdots$$

converges to 0 we must find an N with

$$-\varepsilon < \frac{1}{\log_{10}(N+1)} < \varepsilon \quad \text{and} \quad -\varepsilon < \frac{1}{\log_{10}(N+2)} < \varepsilon \quad \text{and}$$

$$-\varepsilon < \frac{1}{\log_{10}(N+3)} < \varepsilon \quad \text{and} \quad \ldots$$

This will be true provided that $\log_{10}(N+1) > 1/\varepsilon$. So let N be any integer larger than $10^{1/\varepsilon}$. Then

$$\log_{10}(N+1) > \log_{10}(10^{1/\varepsilon}) = 1/\varepsilon \quad \text{and}$$

$$0 < \cdots < \frac{1}{\log_{10}(N+2)} < \frac{1}{\log_{10}(N+1)} < \varepsilon$$

It follows that

$$-\varepsilon < \frac{1}{\log_{10}(N+1)} < \varepsilon \quad \text{and} \quad -\varepsilon < \frac{1}{\log_{10}(N+2)} < \varepsilon \quad \text{and}$$

$$-\varepsilon < \frac{1}{\log_{10}(N + 3)} < \varepsilon \quad \text{and} \quad \dots$$

(again with the left-hand inequality trivially satisfied in each case). Therefore for any $\varepsilon > 0$ we can find an N as required and it follows from the definition that the sequence converges to 0.

(Checking that

$$x - \varepsilon < x_N < x + \varepsilon \quad \text{and} \quad x - \varepsilon < x_{N+1} < x + \varepsilon \quad \text{and} \quad \dots$$

is a great waste of effort. Now you've digested the meaning of the definition of convergence to the limit x there is an easier way of combining all those statements. Given $\varepsilon > 0$ you have to find an integer N such that

$$x - \varepsilon < x_n < x + \varepsilon \quad \textbf{for all } n \geqslant N$$

For example in this last exercise given $\varepsilon > 0$ we let N be any integer larger than $10^{1/\varepsilon}$. Then **for any** $n \geqslant N$ we have

$$\log_{10}(n + 1) \geqslant \log_{10}(N + 1) > 1/\varepsilon \quad \text{and} \quad -\varepsilon < x_n < \varepsilon)$$

3

$x_1, x_2, x_3, \dots \to x$	$x_1 - x, x_2 - x, x_3 - x, \dots \to 0$
means that given $\varepsilon > 0$ there is an integer N with	means that given $\varepsilon > 0$ there is an integer N with
$$x - \varepsilon < x_n < x + \varepsilon$$	$$0 - \varepsilon < (x_n - x) < 0 + \varepsilon$$
for all $n \geqslant N$.	for all $n \geqslant N$.

But you can add or subtract a number right through an inequality without upsetting it. It is easy to see that subtracting x from the inequalities on the left-hand side gives those inequalities on the right; and adding x to the inequalities on the right-hand side gives those inequalities on the left. Hence the inequalities on both sides are saying the same thing and we can deduce that $x_1, x_2, x_3, \dots \to x$ if and only if $x_1 - x, x_2 - x, x_3 - x, \dots \to 0$ as required.

Therefore by the result applied twice we see that

$$x_1, \quad x_2, \quad x_3, \dots \to x$$

$$\Leftrightarrow x_1 - x, \quad x_2 - x, \quad x_3 - x, \dots \to 0$$

'if and only if'

$$\Leftrightarrow (x_1 + y) - (x + y), \quad (x_2 + y) - (x + y), \quad (x_3 + y) - (x + y), \dots \to 0$$

$$\Leftrightarrow x_1 + y, \quad x_2 + y, \quad x_3 + y, \dots \to x + y$$

as required.

4

$x_1, x_2, x_3, \dots \to x$	$-x_1, -x_2, -x_3, \dots \to -x$
means that given $\varepsilon > 0$ there is an integer N with	means that given $\varepsilon > 0$ there is an integer N with
$$x - \varepsilon < x_n < x + \varepsilon$$	$$(-x) - \varepsilon < (-x_n) < (-x) + \varepsilon$$
for all $n \geqslant N$.	for all $n \geqslant N$.

But multiplying through inequalities by -1 completely reverses those inequalities (for example $6 < 7$ but $-6 > -7$). Therefore it is easy to check that multiplying the inequalities on the left-hand side of the page by -1 gives those inequalities on the right (and vice-versa). Hence both sides are saying precisely the same thing and we can deduce that $x_1, x_2, x_3, \ldots \to x$ and only if $-x_1, -x_2, -x_3, \ldots \to -x$ as required.

5    $\quad x_1, x_2, x_3, \ldots \to 0$ | $|x_1|, |x_2|, |x_3|, \ldots \to 0$

means that given $\varepsilon > 0$ there is an integer N with

$$0 - \varepsilon < x_n < 0 + \varepsilon$$

(and, multiplying by -1,

$$0 + \varepsilon > -x_n > 0 - \varepsilon)$$

for all $n \geqslant N$.

means that given $\varepsilon > 0$ there is an integer N with

$$0 - \varepsilon < |x_n| < 0 + \varepsilon$$

(and, multiplying by -1,

$$0 + \varepsilon > -|x_n| > 0 - \varepsilon)$$

for all $n \geqslant N$.

But since $x_n = \pm |x_n|$ and $|x_n| = \pm x_n$ it is clear that both sides are saying the same thing. Hence $x_1, x_2, x_3, \ldots \to 0$ if and only if $|x_1|, |x_2|, |x_3|, \ldots \to 0$.

The result is not true for sequences with non-zero limits. For example the sequence $1, -1, 1, -1, 1, \ldots$ is not convergent whereas the sequence $|1|, |-1|, |1|, |-1|, 1, \ldots$ is clearly convergent to 1.

6   The sequence $x_1, x_2, x_3, \ldots$ converges to 0. Therefore given any $\varepsilon > 0$ there exists an integer N with

$$-\varepsilon < x_n < \varepsilon \quad \text{for all } n \geqslant N$$

(and hence also

$$-\varepsilon < -x_n < \varepsilon \quad \text{for all } n \geqslant N)$$

But $-x_n \leqslant y_n \leqslant x_n$ for all n and so

$$-\varepsilon < -x_n \leqslant y_n \leqslant x_n < \varepsilon \quad \text{for all } n \geqslant N$$

Hence we also have

$$-\varepsilon < y_n < \varepsilon \quad \text{for all } n \geqslant N$$

and it follows that the sequence $y_1, y_2, y_3, \ldots$ also converges to 0.

Page 50

1   Let $x_1, x_2, x_3, \ldots \to x$ and consider a subsequence $x_{k_1}, x_{k_2}, x_{k_3}, \ldots$. (Then in particular $k_n \geqslant n$ for each n.) Given any $\varepsilon > 0$ there exists an N with

$$x - \varepsilon < x_n < x + \varepsilon \quad \text{for all } n \geqslant N$$

Since $k_N, k_{N+1}, k_{N+2}, \ldots$ are all at least N it follows that

$$x - \varepsilon < x_{k_n} < x + \varepsilon \quad \text{for all } n \geqslant N$$

and that the subsequence also converges to x.

Hence if the sequence

$$1, \tfrac{1}{2}, \tfrac{1}{3}, 1, \tfrac{1}{4}, \tfrac{1}{5}, 1, \tfrac{1}{6}, \tfrac{1}{7}, 1, \ldots$$

converges to some limit x then the subsequences

$$1, 1, 1, 1, 1, \ldots \quad \text{and} \quad \tfrac{1}{2}, \tfrac{1}{3}, \tfrac{1}{4}, \tfrac{1}{5}, \tfrac{1}{6}, \ldots$$

both converge to x. But this is a blatent contradiction since these sequences converge to different limits, namely 1 and 0. Hence the original sequence does not converge.

2   This solution consists of showing that if the even terms of the sequence all get close to x and the odd terms all get close to x then *all* the terms get close to x!

Assume that we are giving some $\varepsilon > 0$. Then

$x_2, x_4, x_6, \ldots \to x$	$x_1, x_3, x_5, \ldots \to x$
means that given $\varepsilon > 0$ there is an integer N_1 with	means that given $\varepsilon > 0$ there is an integer N_2 with
$x - \varepsilon < x_{2m} < x + \varepsilon$	$x - \varepsilon < x_{2m+1} < x + \varepsilon$
for all $2m \geqslant N_1$.	for all $2m + 1 \geqslant N_2$.

Therefore we have

$$x - \varepsilon < x_n < x + \varepsilon \quad \text{if } n = 2m \geqslant N_1 \quad \text{or if } n = 2m + 1 \geqslant N_2$$

Hence if N is the larger of N_1 and N_2 it follows that

$$x - \varepsilon < x_n < x + \varepsilon \quad \text{for all } n \geqslant N$$

and $x_1, x_2, x_3, \ldots \to x$ as required.

3   We could imitate the proof of the theorem that an increasing sequence which is bounded above converges (replacing 'increasing' by 'decreasing', 'above' by 'below' etc) but instead we consider the sequence $-x_1, -x_2, -x_3, \ldots$ which is **increasing** and bounded **above**. So by the theorem the sequence $-x_1, -x_2, -x_3, \ldots$ converges. But then it follows from exercise 4 on page 42 that $x_1, x_2, x_3, \ldots$ also converges.

4   Let $x_1, x_2, x_3, \ldots \to x$, where each x_n is non-negative. Assume that $x < 0$: (we shall deduce a contradiction). Since x is negative the number $-x$ is positive and we may take ε to be $-x$ in the definition of the convergence of $x_1, x_2, x_3, \ldots$. In particular there exists an N with

$$x - \varepsilon = x - (-x) < x_N < x + \varepsilon = x + (-x) = 0$$

But then $x_N < 0$ which contradicts the fact that all the terms of the sequence are non-negative. This contradiction shows that $x \geqslant 0$ as required.

Therefore by exercise 3 on page 42

$$y_1, y_2, y_3, \ldots \to y \quad \text{with each } y_n \geqslant a$$
$$\Rightarrow y_1 - a, y_2 - a, y_3 - a, \ldots \to y - a \quad \text{with each } y_n - a \geqslant 0$$

and by what we have just proved it follows that $y - a \geqslant 0$ and $y \geqslant a$ as claimed.

Similarly by exercise 4 on page 42

$$z_1, z_2, z_3, \ldots \to z \quad \text{with each } z_n \leqslant b$$
$$\Rightarrow -z_1, -z_2, -z_3, \ldots \to -z \quad \text{with each } -z_n \geqslant -b$$

and by the previous part it follows that $-z \geqslant -b$ and $z \leqslant b$ as required. Now by the theorem on page 48 any bounded sequence has a convergent subsequence: in particular any sequence in the bounded set $[a, b]$ has a convergent subsequence. All its terms x_{k_n} satisfy $x_{k_n} \geq a$ and $x_{k_n} \leq b$ and so by the previous parts of this exercise the limit also satisfies these inequalities. Hence the convergent subsequence has its limit in $[a, b]$ as required.

5   The difference between $\frac{1}{6}\pi$ and $\frac{5}{6}\pi$ is $\frac{2}{3}\pi$ which is more than 1 and so there is an integer k_1 between $\frac{1}{6}\pi$ and $\frac{5}{6}\pi$ as claimed. Similarly there exist $k_2, k_3, \ldots$ and $j_1, j_2, j_3 \ldots$ as stated. But then for each n we have $\sin k_n > \frac{1}{2}$ and $\sin j_n < -\frac{1}{2}$.

Assume for the moment that the sequence $\sin 1, \sin 2, \sin 3, \ldots$ converges to some number x (we shall deduce a contradiction). By exercise 1 above the subsequence $\sin k_1, \sin k_2, \sin k_3, \ldots$ also converges to x and so by exercise 4 we have that $x \geq \frac{1}{2}$.

But similarly the subsequence $\sin j_1, \sin j_2, \sin j_3, \ldots$ also converges to x and so $x \leq -\frac{1}{2}$.

We cannot have $x \geq \frac{1}{2}$ and $x \leq -\frac{1}{2}$ and so this contradiction shows that the sequence $\sin 1, \sin 2, \sin 3, \ldots$ is divergent.

6   The sequence $x_1, x_2, x_3, \ldots$ converges to the positive number x. Hence by taking $\varepsilon = \frac{1}{2}x$ in the definition of convergence we see that there exists an N with

$$x - \tfrac{1}{2}x < x_n < x + \tfrac{1}{2}x \quad \text{for all } n \geq N$$

In particular $x_n > \frac{1}{2}x \, (>0)$ for all $n \geq N$ and so $1/x_n < 2/x$ for all $n \geq N$. Therefore the set

$$\left\{ \frac{1}{x_N}, \frac{1}{x_{N+1}}, \frac{1}{x_{N+2}}, \ldots \right\}$$

is bounded below by 0 and above by $2/x$. Hence the set

$$\left\{ \frac{1}{x_1}, \frac{1}{x_2}, \frac{1}{x_3}, \ldots \right\} = \underbrace{\left\{ \frac{1}{x_1}, \frac{1}{x_2}, \ldots, \frac{1}{x_{N-1}} \right\}}_{\substack{\text{finite set} \\ \therefore \text{ bounded}}}$$

$$\cup \underbrace{\left\{ \frac{1}{x_N}, \frac{1}{x_{N+1}}, \frac{1}{x_{N+2}}, \ldots \right\}}_{\substack{\text{bounded between 0} \\ \text{and } 2/x}}$$

being the union of two bounded sets is also bounded.

7   (i) From the definition of the positive numbers $x_1, x_2, x_3, \ldots \to 0$ we know that given any $\varepsilon > 0$ there exists an N with

$$0 < x_n < \varepsilon \quad \text{for all } n \geq N$$

In particular

taking $\varepsilon = 1$ shows that there exists N_1 with $x_n < 1$ (and $1/x_n > 1$) for $n \geq N_1$

taking $\varepsilon = \frac{1}{2}$ shows that there exists N_2 with $x_n < \frac{1}{2}$ (and $1/x_n > 2$) for $n \geqslant N_2$

taking $\varepsilon = \frac{1}{3}$ shows that there exists N_3 with $x_n < \frac{1}{3}$ (and $1/x_n > 3$) for $n \geqslant N_3$

$$\vdots$$

In this way, given any huge number k we can make the following deduction:

taking $\varepsilon = 1/k$ shows that there exists N with $x_n < 1/k$ (and $1/x_n > k$) for $n \geqslant N$

Hence given any large k we have found an N as required in the definition to show that the sequence $1/x_1, 1/x_2, 1/x_3, \ldots$ tends to infinity.

(ii) From the definition of $1/x_1, 1/x_2, 1/x_3, \ldots$ tending to infinity we know that given any number k there exists an N with

$$1/x_n > k \quad \text{for all } n \geqslant N$$

In particular

taking $k = 1$ shows that there exists N_1 with $1/x_n > 1$ (and $0 < x_n < 1$) for $n \geqslant N_1$

taking $k = 2$ shows that there exists N_2 with $1/x_n > 2$ (and $0 < x_n < \frac{1}{2}$) for $n \geqslant N_2$

taking $k = 3$ shows that there exists N_3 with $1/x_n > 3$ (and $0 < x_n < \frac{1}{3}$) for $n \geqslant N_3$

$$\vdots$$

In this way, given any tiny positive number ε we can make the following deduction:

taking $k = 1/\varepsilon$ shows that there exists N with $1/x_n > 1/\varepsilon$ (and $0 < x_n < \varepsilon$) for $n \geqslant N$

Hence given any positive ε we have found an N as required in the definition to show that the sequence $x_1, x_2, x_3, \ldots$ of positive terms is converging to 0.

Page 62

1   (i) By our established results on adding, multiplying and dividing convergent sequences we see that

$$\frac{2n^3 + 1}{3n^3 + n + 2} = \frac{2 + \dfrac{1}{n^3}}{3 + \dfrac{1}{n^2} + \dfrac{2}{n^3}} \to \frac{2 + 0}{3 + 0 + 0} = \frac{2}{3} \quad \text{as} \quad n \to \infty$$

(In fact the 10th term is approximately 0.664 and the 100th term approximately 0.66664.)

(ii) $(1 + 1/\sqrt{n})^2 = (1 + 1/\sqrt{n}) \times (1 + 1/\sqrt{n}) \to (1 + 0) \times (1 + 0) = 1$ as $n \to \infty$. (In fact the 100th term is approximately 1.064 and the 100000th term is 1.0201.)

(iii) Here the nth term of the sequence is $(100 + 5^n)^{1/n}$ which equals $5(100/5^n + 1)^{1/n}$. Now

$$\overbrace{5 = 5(0+1)^{1/n} \leqslant 5\left(\frac{100}{5^n}+1\right)^{1/n}}^{=x_n} \leqslant 5(20+1)^{1/n} = 5 \times 21^{1/n}$$

$$\downarrow \qquad\qquad\qquad\qquad\qquad\qquad\qquad\qquad\qquad\qquad\qquad \downarrow$$
$$5 \qquad\qquad \therefore \quad \downarrow \qquad\qquad\qquad\qquad\qquad 5 \times 1 = 5$$
$$5$$

and so by the 'sandwiching' corollary (page 55) the sequence converges to 5. (The 5th term is approximately 5.032 and the 10th approximately 5.000005.)

2   This sequence is defined by the 'recurrence relation'

$$x_1 = 10 \quad \text{and} \quad x_n = \frac{x_{n-1}}{2} + \frac{1}{x_{n-1}} \quad \text{for} \quad n > 1$$

We have already seen in exercise 3 on page 29 that this sequence is decreasing and bounded below (by 1 for example). Therefore (by exercise 3 on page 50) the sequence converges, to x say. Hence from our various results about combining convergent sequences we can deduce that as $n \to \infty$

$$x_{n-1} \to x \quad \text{and} \quad \frac{x_{n-1}}{2} \to \frac{x}{2} \quad \text{and} \quad \frac{1}{x_{n-1}} \to \frac{1}{x} \quad \text{and}$$

$$\underbrace{\frac{x_{n-1}}{2} + \frac{1}{x_{n-1}}}_{=x_n} \to \frac{x}{2} + \frac{1}{x}$$

But $x_n \to x$ and so we see that

$$x = \frac{x}{2} + \frac{1}{x}$$

from which it is easy to deduce that x is $\sqrt{2}$ as expected.

3   The first few turns of the sequence $(n^{1/n})$ are approximately

$$1, 1.4142, 1.4422, 1.4142, 1.3797, 1.3480, \ldots$$

so they soon seem to be decreasing (towards 1?). The 100th term is approximately 1.0471 and the 1000th approximately 1.0069.

   To see formally that the sequence converges to 1 we note from Bernoulli's inequality with $x = 1/\sqrt{n}$ that

$$1 + \sqrt{n} = 1 + nx \leqslant (1+x)^n = \left(1 + \frac{1}{\sqrt{n}}\right)^n$$

Raising each side of the inequality to the power $2/n$ (i.e. squaring and taking the nth roots) gives

$$(1 + \sqrt{n})^{2/n} \leqslant \left(1 + \frac{1}{\sqrt{n}}\right)^2$$

and hence

$$1 \leqslant n^{1/n} = (\sqrt{n})^{2/n} < (1 + \sqrt{n})^{2/n} \leqslant \left(1 + \frac{1}{\sqrt{n}}\right)^2$$
$$\downarrow \quad \downarrow \qquad\qquad\qquad\qquad\qquad\qquad\qquad\qquad \downarrow$$
$$1 \quad \downarrow \qquad\qquad\qquad\qquad\qquad\qquad\qquad\qquad 1$$
$$\therefore \ 1$$

Therefore by sandwiching we see that $n^{1/n} \to 1$ as $n \to \infty$.

4   In this sequence $x_n = x^n/n!$ and so

$$x_{n+1} \div x_n = \frac{x^{n+1}}{(n+1)!} \div \frac{x^n}{n!} = \frac{x}{n+1} < 1 \quad \text{for } n \geqslant N \ (>x)$$

Hence the sequence $x_N, x_{N+1}, x_{N+2}, \ldots$ is decreasing and bounded below (by 0) and so the sequence (x_n) converges, to α say. But then as $n \to \infty$

$$x_{n+1} = \frac{x^{n+1}}{(n+1)!} = x \times \frac{1}{n+1} \times \frac{x^n}{n!} = x \times \frac{1}{n+1} \times x_n \to x \times 0 \times \alpha = 0$$

Since x_{n+1} also converges to α it follows that $\alpha = 0$ as required.

Page 73

1   (i) $0 \leqslant a_n = \dfrac{(n^2+1)^3}{(n^4+1)^2} \leqslant \dfrac{(2n^2)^3}{(n^4)^2} = \dfrac{8}{n^2}$

But the series $\sum 1/n^2$ converges and hence so does the series $\sum 8/n^2$. Therefore by the comparison test (page 66) the series $\sum a_n$ converges.

(ii) $\dfrac{a_{n+1}}{a_n} = \dfrac{5^{2n+2}((n+1)!)^3}{(3n+3)!} \div \dfrac{5^{2n}(n!)^3}{(3n)!} = 25 \dfrac{(n+1)(n+1)(n+1)}{(3n+3)(3n+2)(3n+1)}$

$$= 25 \frac{(1+1/n)(1+1/n)(1+1/n)}{(3+3/n)(3+2/n)(3+1/n)} \to \frac{25}{27} \quad \text{as} \quad n \to \infty$$

Hence by the ratio test (page 69) the series $\sum a_n$ converges.

(iii) We know from exercise 5 on page 51 that the sequence $(\sin n)$ does not converge. Therefore the separate terms of the series $\sum \sin n$ do not tend to 0 and, by the theorem on page 65, the series diverges.

2   Let s_n be the nth partial sum of the series $\sum a_n$ and let t_n be the nth partial sum of the series $\sum b_n$. Then it is easy to check that for $n \geqslant N$

$$t_n = s_n + t \to s + t \quad \text{as} \quad n \to \infty$$

and so the series $\sum b_n$ converges with sum $s + t$.

(Hence you can change any finite number of terms of a series without upsetting its convergence.)

3   If $r \leqslant 1$ then the terms of the series $\sum 1/n^r$ are at least as big as those of $\sum 1/n$, which diverges. It follows that the series $\sum 1/n^r$ diverges. (Formally if the series $\sum 1/n^r$ converged you could use the comparison test with

$$0 \leqslant \frac{1}{n} \leqslant \frac{1}{n^r}$$

to deduce that $\sum 1/n$ converged, which is a contradiction.)

Now if $r > 1$ then from exercise 3 on page 22 we see that the partial sums of the series $\sum 1/n^r$ are bounded above (by $1/(1 - (\frac{1}{2})^{r-1})$). Hence the sequence of partial sums of this series of positive terms is increasing and bounded above. It follows that the series $\sum 1/n^r$ converges in this case.

It is interesting to note therefore that although the series

$$1 + \tfrac{1}{2} + \tfrac{1}{3} + \tfrac{1}{4} + \cdots$$

diverges, the series

$$1 + \frac{1}{2^{1.0001}} + \frac{1}{3^{1.0001}} + \frac{1}{4^{1.0001}} + \frac{1}{5^{1.0001}} + \cdots$$

converges (with sum less than or equal to $1/(1 - (\frac{1}{2})^{0.0001}) \approx 14\ 428$).

4   If $s_2, s_4, s_6, \ldots \to s$ and $a_1, a_2, a_3, \ldots$ converges to 0 then

$$s_{2n+1} = s_{2n} + a_{2n+1} \to s + 0 = s \quad \text{as} \quad n \to \infty$$

Hence the sequence $s_1, s_3, s_5, \ldots$ also converges to s and we can deduce from exercise 2 on page 50 that the sequence (s_n) converges to s.

(That earlier exercise 2 can easily be extended to the case where the sequence (s_n) is broken down into M sequences each of which converges to s. Then if the subsequence (s_{Mn}) of partial sums converges to s and the sequence (a_n) converges to 0 it is easy to deduce as above that all the sequences $(s_{Mn+1}), (s_{Mn+2}), \ldots (s_{Mn+n-1})$ converge to s, and hence that the sequence (s_n) converges to s.)

5        $t_{3n} = 1 - \frac{1}{2} - \frac{1}{4} + \frac{1}{3} - \frac{1}{6} - \frac{1}{8} + \cdots + \dfrac{1}{2n-1} - \dfrac{1}{4n-2} - \dfrac{1}{4n}$

$$= \left(1 + \tfrac{1}{2} + \tfrac{1}{3} + \cdots + \frac{1}{2n}\right) - \left(\tfrac{1}{2} + \tfrac{1}{4} + \cdots + \frac{1}{2n}\right)$$

$$\qquad - \left(\tfrac{1}{2} + \tfrac{1}{4} + \cdots + \frac{1}{4n}\right)$$

$$= \left(1 + \tfrac{1}{2} + \tfrac{1}{3} + \cdots + \frac{1}{2n}\right) - \left(\tfrac{1}{2} + \tfrac{1}{4} + \cdots + \frac{1}{2n}\right)$$

$$\qquad - \tfrac{1}{2}\left(1 + \tfrac{1}{2} + \cdots + \frac{1}{2n}\right)$$

$$= \tfrac{1}{2}\left(1 + \tfrac{1}{2} + \tfrac{1}{3} + \cdots + \frac{1}{2n}\right) - \left(\tfrac{1}{2} + \tfrac{1}{4} + \cdots + \frac{1}{2n}\right)$$

$$= \tfrac{1}{2}\left(1 - \tfrac{1}{2} + \tfrac{1}{3} - \tfrac{1}{4} + - \cdots - \frac{1}{2n}\right)$$

$$= \tfrac{1}{2}s_{2n} \to \tfrac{1}{2}s \quad \text{as} \quad n \to \infty$$

Hence the subsequence (t_{3n}) of partial sums converges to $\frac{1}{2}s$ and it follows from the previous exercise that the rearranged series converges with sum $\frac{1}{2}s$.

6   (i) We have seen (on page 47) that

$$1 + \tfrac{1}{2} + \tfrac{1}{3} + \cdots + \frac{1}{n-1} - \log n \to \gamma \quad \text{as} \quad n \to \infty$$

Hence as $n \to \infty$ we have

$$1 + \tfrac{1}{2} + \tfrac{1}{3} + \cdots + \frac{1}{n} - \log n = \left(1 + \tfrac{1}{2} + \tfrac{1}{3} + \cdots + \frac{1}{n-1} - \log n\right) + \frac{1}{n}$$

$$\to \gamma + 0 = \gamma$$

i.e. the sequence $(u_n - \log n)$ converges to Euler's constant γ.

(ii) Now, rather as in the previous solution,

$$s_{2n} = 1 - \tfrac{1}{2} + \tfrac{1}{3} - \tfrac{1}{4} + \cdots - \frac{1}{2n}$$

$$= \left(1 + \tfrac{1}{2} + \tfrac{1}{3} + \cdots + \frac{1}{2n}\right) - 2\left(\tfrac{1}{2} + \tfrac{1}{4} + \tfrac{1}{6} + \cdots + \frac{1}{2n}\right)$$

$$= \left(1 + \tfrac{1}{2} + \tfrac{1}{3} + \cdots + \frac{1}{2n}\right) - \left(1 + \tfrac{1}{2} + \tfrac{1}{3} + \cdots + \frac{1}{n}\right)$$

$$= u_{2n} - u_n$$

$$= (u_{2n} - \log 2n) - (u_n - \log n) + \log 2n - \log n$$

$$= (u_{2n} - \log 2n) - (u_n - \log n) + \underbrace{\log(2n/n)}_{= \log 2}$$

$$\to \gamma - \gamma + \log 2 = \log 2$$

(iii) Hence the partial sums $s_2, s_4, s_6, \ldots$ of the series $1 - \tfrac{1}{2} + \tfrac{1}{3} - \tfrac{1}{4} + \cdots$ converge to $\log 2$ and it follows from exercise 4 that the series converges with sum $\log 2$.

In fact my computer gives the 1000th partial sum of this series as 0.6936463 (and, fitting in with the result of the previous exercise, it gives the 1500th partial sum of the rearranged series as 0.3468232).

7   For the series $\sum |x^n|$ the ratio of successive terms is given by

$$\frac{(n+1)\text{th term}}{n\text{th term}} = \frac{|x^{n+1}|}{|x^n|} = |x|$$

Hence by the ratio test (page 69) for $|x| < 1$ the series $\sum |x^n|$ is convergent; i.e. the series $\sum x^n$ is absolutely convergent (and hence convergent) for $-1 < x < 1$.

If $|x| \geqslant 1$ then the terms of the series do not converge to 0 and so the series diverges. Hence D is the interval $(-1, 1)$.

This particular series is a geometric series and by summing it you can see that $f(x) = 1/(1 - x)$ for $x \in (-1, 1)$.

Chapter 3

Page 79

1   (i) $f(x) = x^2$ $(x \in \mathbb{R})$ is not one-to-one because, for example,

$$f(2) = f(-2) = 4$$

(ii) The function $g(x) = 1/x$ $(x \in \mathbb{R}, x \neq 0)$ has domain and range consisting of the non-zero numbers and it is one-to-one. For given any y in the range we can find a unique x with $g(x) = y$:

$$y = g(x) = \frac{1}{x} \quad \Leftrightarrow \quad x = \frac{1}{y}$$

Hence $g^{-1}(y) = 1/y$ and g is its own inverse. For example

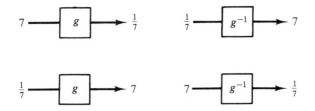

(iii) To see whether $h(x) = (2x - 1)/(x + 1)$ $(x \in \mathbb{R}, x \neq -1)$ is one-to-one consider the equation

$$y = h(x) = \frac{2x - 1}{x + 1} \quad (x \neq -1)$$

This can be uniquely solved for x to give

$$xy + y = 2x - 1 \quad \text{and} \quad x = h^{-1}(y) = \frac{1 + y}{2 - y} \quad (y \neq 2)$$

Hence h has domain D consisting of all real numbers apart from -1 and range G consisting of all real numbers apart from 2, and h^{-1} has domain G and range D. The graphs of h and h^{-1} are illustrated below.

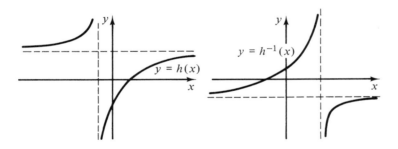

An alternative way of finding h^{-1} is to use a standard school method and break h down into some simpler component parts. In this case

$$h(x) = \frac{2x - 1}{x + 1} = \frac{2x + 2 - 3}{x + 1} = 2 - \frac{3}{x + 1}$$

This can be represented diagrammatically by

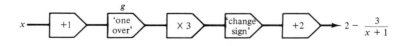

Reversing the process then gives

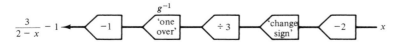

and hence the inverse function is given by

$$h^{-1}(x) = \frac{3}{2-x} - 1 = \frac{1+x}{2-x} \qquad x \in \mathbb{R}, x \neq 2$$

Note that in (ii) g was its own inverse: we saw that again when reversing the component parts of h above. We also saw there that the function 'change signs' (i.e. $j(x) = -x$) is its own inverse. The graphs of such functions will be symmetrical about the line $y = x$. Other one-to-one functions with that property include

$$p(x) = \frac{2-x}{1-x} \quad x \in \mathbb{R}, x \neq 1 \quad \text{and} \quad q(x) = \begin{cases} -2x & x \leqslant 0 \\ -\frac{1}{2}x & x > 0 \text{ etc.} \end{cases}$$

The symmetrical graphs of g, j, p and q are illustrated below.

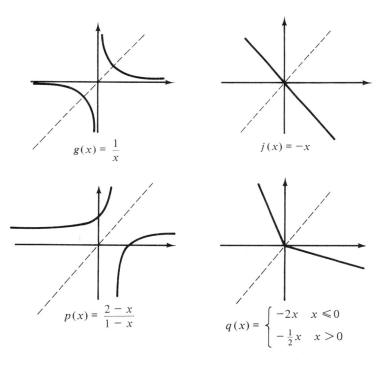

$$g(x) = \frac{1}{x}$$

$$j(x) = -x$$

$$p(x) = \frac{2-x}{1-x}$$

$$q(x) = \begin{cases} -2x & x \leqslant 0 \\ -\frac{1}{2}x & x > 0 \end{cases}$$

2   If f is strictly increasing then $x_1 < x_2$ (in f's domain) implies that $f(x_1) < f(x_2)$. Hence it is impossible for $f(x_1)$ to equal $f(x_2)$ for different x_1 and x_3. Thus each y in f's range equals $f(x)$ for some *unique* x in its domain, and f is one-to-one. A similar argument works in the decreasing case.

     The one-to-one function g in exercise 1 is neither increasing nor decreasing.

3   (i) For fixed x_1, x_2 and $\alpha, \beta \geqslant 0$ with $\alpha + \beta = 1$ the quadratic in T given by

$$\underbrace{(\alpha + \beta)}_{= 1}T^2 + 2(\alpha x_1 + \beta x_2)T + (\alpha x_1^2 + \beta x_2^2) = \alpha(T + x_1)^2 + \beta(T + x_2)^2 \geqslant 0$$

is never negative. But if a quadratic $aT^2 + bT + c$ is never negative we know that its 'discriminant' is zero or less; i.e. $b^2 - 4ac \leqslant 0$ (otherwise the formula for the solution of the quadratic equation would yield two roots and this quadratic – with a single T^2 – would be negative between those roots). Hence the discriminant of the displayed quadratic is zero or less and we deduce that

$$(2(\alpha x_1 + \beta x_2))^2 - 4 \times 1 \times (\alpha x_1{}^2 + \beta x_2{}^2) \leqslant 0$$

and

$$(\alpha x_1 + \beta x_2)^2 \leqslant \alpha x_1{}^2 + \beta x_2{}^2$$

as required.

(ii) To show that the function f is convex we must take any numbers x_1 and x_2 in f's domain and show that for any number z between them $f(z)$ is less than or equal to the height h of the chord, as illustrated.

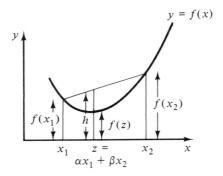

As we've seen before (in the solution of exercise 5 on pages 226 and 227 for example) any number between x_1 and x_2 is of the form $z = \alpha x_1 + \beta x_2$ for some non-negative α and β with $\alpha + \beta = 1$.

For such a z we have the height h of the chord is

$$\alpha f(x_1) + \beta f(x_2)$$

Hence a function is convex if and only if for each x_1 and x_2 in its domain and each $\alpha, \beta \geqslant 0$ with $\alpha + \beta = 1$ we have

$$f(z) = f(\alpha x_1 + \beta x_2) \leqslant \alpha f(x_1) + \beta f(x_2)$$

(iii) If the function g is given by

$$g(x) = x^2 \quad x \in \mathbb{R}$$

then by (ii) g will be convex if and only if for each $x_1, x_2 \in \mathbb{R}$ and each $\alpha, \beta \geqslant 0$ with $\alpha + \beta = 1$ we have

$$g(\alpha x_1 + \beta x_2) \leqslant \alpha g(x_1) + \beta g(x_2)$$
$$\| \qquad\qquad \| \qquad\qquad \|$$
$$(\alpha x_1 + \beta x_2)^2 \quad \alpha x_1{}^2 \quad \beta x_2{}^2$$

But that is precisely the inequality which we established in (i) and hence the function g is convex.

(iv) If we restrict the domain of $g(x) = x^2$ to the non-negative numbers then, as we have seen, g has an inverse function given by $h(x) = g^{-1}(x) = \sqrt{x}$ $(x \geqslant 0)$. Since g is convex the chords of its graph always lie on or above the graph as illustrated below. The graph of g^{-1} is obtained from that of g by reflecting it in the line $y = x$ and it seems that all the chords of the graph of g^{-1} will lie on or below that graph; i.e. that g^{-1} is concave.

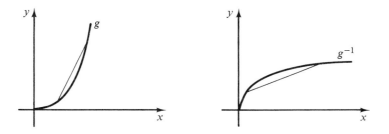

To prove formally that g^{-1} is concave we note that, rather as in (ii) above, g^{-1} is concave if and only if for each $x_1, x_2 \geqslant 0$ and each $\alpha, \beta \geqslant 0$ with $\alpha + \beta = 1$ we have

$$g^{-1}(\alpha x_1 + \beta x_2) \geqslant \alpha g^{-1}(x_1) + \beta g^{-1}(x_2)$$
$$\| \qquad\qquad \| \qquad\quad \|$$
$$\sqrt{(\alpha x_1 + \beta x_2)} \qquad \alpha\sqrt{x_1} \qquad \beta\sqrt{x_2}$$

But by the inequality established in (i) applied now to the numbers $\sqrt{x_1}$ and $\sqrt{x_2}$ we see that

$$\alpha x_1 + \beta x_2 = \alpha(\sqrt{x_1})^2 + \beta(\sqrt{x_2})^2 \geqslant (\alpha\sqrt{x_1} + \beta\sqrt{x_2})^2$$

and taking non-negative square roots enables us to deduce that

$$g^{-1}(\alpha x_1 + \beta x_2) \geqslant \alpha g^{-1}(x_1) + \beta g^{-1}(x_2)$$

and that g^{-1} is concave, as required.

(Keen readers might like to consider any strictly increasing convex function f – whose domain and range will both be intervals – and show by a very similar deduction that the inverse function f^{-1} is concave.)

Page 88

1   Let x and y be any numbers and let (x_n) and (y_n) be sequences of rationals converging to x and y respectively; i.e.

$$x_1, x_2, x_3, \ldots \to x \qquad y_1, y_2, y_3, \ldots \to y$$

Then by our result about adding and subtracting convergent sequences we have

$$x_1 + y_1, x_2 + y_2, x_3 + y_3, \ldots \to x + y$$

and

$$x_1 - y_1, x_2 - y_2, x_3 - y_3, \ldots \to x - y$$

Hence by our derived property of the function 10^x (or, in more precise parlance, the function whose value at x is 10^x) and in general of a^x we have

$$a^{x_1}, a^{x_2}, a^{x_3}, \ldots \to a^x \qquad a^{y_1}, a^{y_2}, a^{y_3}, \ldots \to a^y$$

and

$$a^{x_1 + y_1}, a^{x_2 + y_2}, a^{x_3 + y_3}, \ldots \to a^{x+y}$$

and

$$a^{x_1 - y_1}, a^{x_2 - y_2}, a^{x_3 - y_3}, \ldots \to a^{x-y}$$

Now from the result about multiplying/dividing convergent sequences and from the fact that $a^{x_n} \times a^{y_n} = a^{x_n + y_n}$ and $a^{x_n} \div a^{y_n} = a^{x_n - y_n}$ for the rationals x_n and y_n we can deduce that

$$\left. \begin{array}{c} a^{x_1} \times a^{y_1}, a^{x_2} \times a^{y_2}, a^{x_3} \times a^{y_3}, \ldots \to a^x \times a^y \\ \| \qquad\quad \| \qquad\quad \| \\ a^{x_1 + y_1}, \qquad a^{x_2 + y_2}, \qquad a^{x_3 + y_3}, \quad \ldots \to a^{x+y} \end{array} \right\} \therefore \ a^x \times a^y = a^{x+y}$$

and

$$\left. \begin{array}{c} a^{x_1} \div a^{y_1}, a^{x_2} \div a^{y_2}, a^{x_3} \div a^{y_3}, \ldots \to a^x \div a^y \\ \| \qquad\quad \| \qquad\quad \| \\ a^{x_1 - y_1}, \qquad a^{x_2 - y_2}, \qquad a^{x_3 - y_3}, \quad \ldots \to a^{x-y} \end{array} \right\} \therefore \ a^x \div a^y = a^{x-y}$$

as required.

(And in general we can assume henceforth that powers behave exactly as we'd expect.)

To show that a^x is strictly increasing for $a > 1$ note that if $x < y$ then $y - x > 0$ and $a^{y-x} > 1$. But then

$$\frac{a^y}{a^x} = a^{y-x} > 1 \quad \text{and so} \quad a^x < a^y$$

and the function is increasing. In the case where $a < 1$ these last two inequalities are reversed and the function is therefore strictly decreasing.

2 Note firstly that $f(0) = 1$ since

$$f(1) = f(0 + 1) = f(0) \times \underbrace{f(1)}_{>0} \quad \text{and so} \quad f(0) = \frac{f(1)}{f(1)} = 1$$

Next note that for any positive integer n

$$f(n) = f(\underbrace{1 + 1 + \cdots + 1}_{\leftarrow \ n \text{ times} \ \rightarrow}) = \underbrace{f(1) \times f(1) \times \cdots \times f(1)}_{\longleftarrow \ n \text{ times} \ \longrightarrow} = (f(1))^n = a^n$$

(or, if you're fussy, this result can be proved by induction). Therefore for that same positive integer n we have

$$f(-n) = \frac{f(n) \times f(-n)}{f(n)} = \frac{f(n + (-n))}{f(n)} = \frac{f(0)}{f(n)} = \frac{1}{a^n} = a^{-n}$$

Hence we have shown that $f(x) = a^x$ for each *integer* x. To extend that result to rational x note that for integers m and n with $n > 0$ we have

$$\left(f\left(\frac{m}{n}\right)\right)^n = \underbrace{f\left(\frac{m}{n}\right) \times f\left(\frac{m}{n}\right) \times \cdots \times f\left(\frac{m}{n}\right)}_{n \text{ times}}$$

$$= f\left(\underbrace{\frac{m}{n} + \frac{m}{n} + \cdots + \frac{m}{n}}_{n \text{ times}}\right) = f(m) = a^m$$

and so

$$f\left(\frac{m}{n}\right) = \sqrt[n]{(a^m)} = a^{m/n}$$

and $f(x) = a^x$ for any *rational* x.

That is as far as we can go by these finite/algebraic/primary-school methods. To extend the result to *all* real numbers we need the given additional analytical property concerning sequences. So now given any number x let (x_n) be a sequence of rationals which converges to x. Then by the given property of f and by the fact that $f(x_n) = a^{x_n}$ for each rational x_n we can see that

$$\left.\begin{array}{c} x_1, \quad x_2, \quad x_3, \ldots \to x \\ \therefore \quad f(x_1), f(x_2), f(x_3), \ldots \to f(x) \\ \| \quad \quad \| \quad \quad \| \\ a^{x_1}, \quad a^{x_2}, \quad a^{x_3}, \ldots \to a^x \end{array}\right\} \therefore f(x) = a^x$$

and hence $f(x) = a^x$ for this arbitrary x, as required.

3 Let $\alpha = \log_a x$ and $\beta = \log_a y$. Then $a^\alpha = x$ and $a^\beta = y$ and so

$$a^{\alpha+\beta} = a^\alpha a^\beta = xy \quad \text{and} \quad a^{\alpha-\beta} = \frac{a^\alpha}{a^\beta} = \frac{x}{y}$$

Hence

$$\log_a (xy) = \alpha + \beta = \log_a x + \log_a y$$

and

$$\log_a\left(\frac{x}{y}\right) = \alpha - \beta = \log_a x - \log_a y$$

4 In the example on page 86 we used the method of log tables to calculate the value of 2.69×7.39 as approximately 19.9 by adding their logs. Using the logarithmic rulers this would be equivalent to

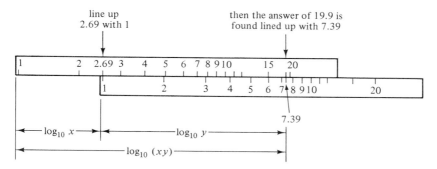

For division we would use the relationship $\log_{10}(x/y) = \log_{10} x - \log_{10} y$ established above. For example to calculate $23.5 \div 5.71$ we would line up the rulers as shown:

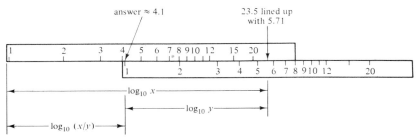

That is the basic principle of the 'slide rule'. If one of the rulers had in addition a 'loglog' scale (so that alongside the mark for a number x the number $\log x$ was also noted) can you see how the slide rule could be used to calculate x^y?

Page 95

1   The graphs of the functions will enable us to guess the nature of the three limits and then we'll give the formal verifications.

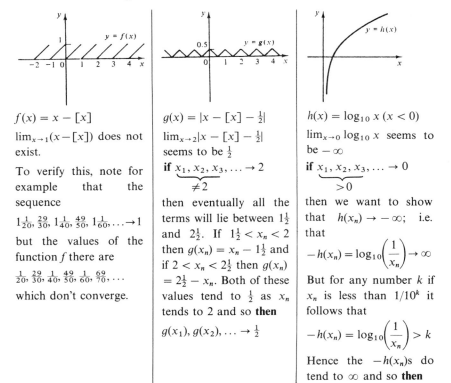

$f(x) = x - [x]$

$\lim_{x \to 1}(x - [x])$ does not exist.

To verify this, note for example that the sequence

$$1\tfrac{1}{20}, \tfrac{29}{30}, 1\tfrac{1}{40}, \tfrac{49}{50}, 1\tfrac{1}{60}, \ldots \to 1$$

but the values of the function f there are

$$\tfrac{1}{20}, \tfrac{29}{30}, \tfrac{1}{40}, \tfrac{49}{50}, \tfrac{1}{60}, \tfrac{69}{70}, \cdots$$

which don't converge.

$g(x) = |x - [x] - \tfrac{1}{2}|$

$\lim_{x \to 2}|x - [x] - \tfrac{1}{2}|$ seems to be $\tfrac{1}{2}$

if $\underbrace{x_1, x_2, x_3, \ldots \to 2}_{\neq 2}$

then eventually all the terms will lie between $1\tfrac{1}{2}$ and $2\tfrac{1}{2}$. If $1\tfrac{1}{2} < x_n < 2$ then $g(x_n) = x_n - 1\tfrac{1}{2}$ and if $2 < x_n < 2\tfrac{1}{2}$ then $g(x_n) = 2\tfrac{1}{2} - x_n$. Both of these values tend to $\tfrac{1}{2}$ as x_n tends to 2 and so **then**

$$g(x_1), g(x_2), \ldots \to \tfrac{1}{2}$$

$h(x) = \log_{10} x \; (x < 0)$

$\lim_{x \to 0} \log_{10} x$ seems to be $-\infty$

if $\underbrace{x_1, x_2, x_3, \ldots \to 0}_{> 0}$

then we want to show that $h(x_n) \to -\infty$; i.e. that

$$-h(x_n) = \log_{10}\left(\frac{1}{x_n}\right) \to \infty$$

But for any number k if x_n is less than $1/10^k$ it follows that

$$-h(x_n) = \log_{10}\left(\frac{1}{x_n}\right) > k$$

Hence the $-h(x_n)$s do tend to ∞ and so **then** $h(x_1), h(x_2), \ldots \to -\infty$

2   Let us consider first what happens to f as x tends to a number $x_0 \neq 1$:
if $\underbrace{x_1, x_2, x_3, \ldots}_{\neq x_0} \to x_0$

then eventually none of the terms is 1 and so

then $f(x_n) = \dfrac{|x_n|^2 + x_n|x_n| - 2}{x_n - 1} \to \dfrac{|x_0|^2 + x_0|x_0| - 2}{x_0 - 1} = f(x_0)$

Hence $\lim_{x \to x_0} f(x) = f(x_0)$ as required.

We now consider the limit of f as x tends to 1. For $x \geqslant 0$ with $x \neq 1$ we have

$$f(x) = \frac{|x|^2 + x|x| - 2}{x - 1} = \frac{2(x^2 - 1)}{x - 1} = 2(x + 1)$$

Hence

if $\underbrace{x_1, x_2, x_3, \ldots}_{\neq 1} \to 1$

then eventually the terms are positive and so

then $f(x_n) = 2(x_n + 1) \to 4$

Therefore $\lim_{x \to 1} f(x) = 4$ and the graph of f is as shown:

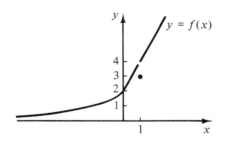

Clearly to avoid that break in the graph of f at $x = 1$ we would have to change the value of $f(1)$ to 4.

Page 106

1   **If** $\underbrace{x_1, x_2, x_3, \ldots}_{} \to x_0$

in $f \circ g$'s domain

then (by the continuity of g at x_0)

$g(x_1), g(x_2), g(x_3), \ldots \to g(x_0)$

and (by the continuity of f at $g(x_0)$)

then $f(g(x_1)), f(g(x_2)), f(g(x_3)), \ldots \to f(g(x_0))$

and so the function $f \circ g$ is continuous at x_0.

We shall now express h as a composite function. So let $f(x) = [x]$ $(x \in \mathbb{R})$ and let $g(x) = \log_2 x (x > 0)$. Then $h(x) = f(g(x))$ and so h is the composite function $f \circ g$.

Now g is the inverse of the strictly increasing function 2^x and by the theorem on page 101 g is continuous at any $x_0 > 0$: in particular g is continuous at $x = 3$ and $g(3) \approx 1.5849$.

Also it is easy to check that f is continuous at each non-integral number: in particular f is continuous at $g(3)$. For

$$\textbf{if} \quad x_1, x_2, x_3, \ldots \underbrace{x_N, x_{N+1}, x_{N+2}, \ldots}_{\text{between 1 and 2}} \to g(3) \, (= 1.5849 \ldots)$$

then the terms are all eventually between 1 and 2 as shown. So

$$\textbf{then} \quad f(x_1), f(x_2), f(x_3), \ldots f(x_N), f(x_{N+1}), f(x_{N+2}), \ldots \to f(g(3))$$

$$\| \qquad \| \qquad \| \qquad \qquad \|$$
$$1 \qquad 1 \qquad 1 \qquad \qquad 1$$

and f is continuous at $g(3)$.

So by the first part of the exercise $f \circ g \, (= h)$ is continuous at $x = 3$.

2   (i) and (ii)   **If** $x_1, x_2, x_3, \ldots \to x_0$

then by multiplying this sequence by itself $n - 1$ times we can deduce that

$$\textbf{then} \quad x_1^{\,n}, \quad x_2^{\,n}, \quad x_3^{\,n}, \ldots \to x_0^{\,n}$$
$$\| \qquad \| \qquad \| \qquad \qquad \|$$
$$f(x_1), \quad f(x_2), \quad f(x_3), \ldots \to f(x_0)$$

and also (provided that x_0 and the x_ns are non-zero to make sure they're in g's domain) we know that

$$\textbf{then} \quad \frac{1}{x_1^{\,n}}, \quad \frac{1}{x_2^{\,n}}, \quad \frac{1}{x_3^{\,n}}, \quad \ldots \to \frac{1}{x_0^{\,n}}$$
$$\| \qquad \| \qquad \| \qquad \qquad \|$$
$$g(x_1), \quad g(x_2), \quad g(x_3), \ldots \to g(x_0)$$

and so f and g are continuous at each x_0 in their domains; i.e. they are continuous.

(iii) Now if we restrict the domain of f to those x with $x \geqslant 0$ then f is strictly increasing. Hence the theorem on page 101 shows that its inverse function is continuous. But the inverse function is precisely the nth root function h in the exercise.

(iv) Assume that r is the rational number m/n where m and n are integers and $n > 0$. Then

$$j(x) = x^r = x^{m/n} = (x^{1/n})^m = (\sqrt[n]{x})^m = k(h(x))$$

where h is as in (iii) above and $k(x) = x^m$ (which is like f or g above, depending upon whether m is positive or negative). So j is a composition of continuous functions and hence by exercise 1 it is continuous.

3   Let $x_1 = a$ and for $n > 1$ let $x_n = f(x_{n-1})$ where f is a continuous function, and assume that the sequence converges; i.e.

$$x_1, x_2, x_3, x_4, \ldots \to x_0$$

Then by the continuity of f we have

$$f(x_1), f(x_2), f(x_3), f(x_4), \ldots \to f(x_0)$$

$$\begin{array}{cccc} \| & \| & \| & \| \\ x_2 & x_3 & x_4 & x_5 \end{array}$$

But this sequence is just a subsequence of the original sequence and so it clearly converges to x_0. It follows that $x_0 = f(x_0)$ as required. Applying the process with $x_1 = 0$ and $f(x) = (x^3 + 2x^2 + 1)/8$ gives (approximately) the first few terms of the sequence as

$$0, \ 0.125 \ (= f(0)), \ 0.1292 \ (\approx f(0.125)), \ 0.12943, \ 0.129459, \ 0.129461, \ldots$$

This sequence seems to be converging to a limit of approximately 0.1295 and, if that is true, that limit will be a root of

$$x = f(x) = \frac{x^3 + 2x^2 + 1}{8} \quad \text{or} \quad x^3 + 2x^2 - 8x + 1 = 0$$

Keen readers can check that the cubic is positive at $x = 0.12945$ and negative at $x = 0.12955$, implying (perhaps) that 0.1295 is, correct to four decimal places, a root of the cubic.

4   Since f is strictly increasing (with range $\mathbb{R}$) it has an inverse function f^{-1} (with domain $\mathbb{R}$). Also since f's domain is an interval it follows from the theorem on page 101 that f^{-1} is continuous.

We now check that f^{-1} satisfies (i)' and (ii)'. The fact that $f^{-1}(1) = a$ is an immediate consequence of the fact that $f(a) = 1$. Now given any x and y it follows that they are in f's range and hence equal to some $f(\alpha)$ and $f(\beta)$ respectively. Hence by (ii) applied to α and β we get

$$f(\alpha \times \beta) = f(\alpha) + f(\beta) = x + y$$

and

$$f^{-1}(x) \times f^{-1}(y) = \alpha \times \beta = f^{-1}(x + y)$$

as required.

So f^{-1} has all the properties needed in exercise 2 on page 88 to show that $f^{-1}(x) = a^x$ for each $x \in \mathbb{R}$. It follows that f, being the inverse of this function, has $f(x) = \log_a x$ for each $x > 0$, as required.

5   If no such interval exists then in *each* interval surrounding x_0 there is some number x with $f(x) \leqslant 0$. In particular

> there exists $x_1 \in (x_0 - 1, x_0 + 1)$ with $f(x_1) \leqslant 0$
> there exists $x_2 \in (x_0 - \frac{1}{2}, x_0 + \frac{1}{2})$ with $f(x_1) \leqslant 0$
> there exists $x_3 \in (x_0 - \frac{1}{3}, x_0 + \frac{1}{3})$ with $f(x_1) \leqslant 0$
> $\vdots$

Clearly $x_1, x_2, x_3, \ldots \to x_0$ and so by the continuity of f we have

$$\begin{array}{cccc} f(x_1), & f(x_2), & f(x_3), \ldots & \to f(x_0) \\ \leqslant 0 & \leqslant 0 & \leqslant 0 & > 0 \end{array}$$

This contradiction shows that an interval *does* exist as required.

Page 118

1   f is a composition of the functions $g(x) = e^x$ and $h(x) = -x$, both of which are continuous: hence f is continuous.

The first few terms of the sequence can be calculated (approximately) as

$$0, 1 (=f(0)), 0.368 (\approx f(1)), 0.692, 0.500, 0.606, 0.545, \ldots$$

with the 19th and 20th terms agreeing to four decimal places at $0.5671\ldots$. By exercise 3 on page 107 if this sequence converges its limit is a root of

$$x = f(x) = e^{-x} \quad \text{or} \quad \log x = -x \quad \text{or} \quad x + \log x = 0$$

The value of $x + \log x$ is negative at $x = 0.5665$ and positive at $x = 0.5675$ thus implying (perhaps) that, correct to three decimal places, 0.567 is a root of the given equation.

In fact the sketch graphs of $y = \log x$ and $y = -x$ show that they intersect in just one point and so it seems that we have found an approximation to the *unique* root of $x + \log x = 0$.

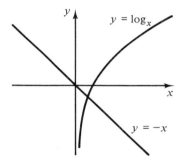

2   (i) $e^{x \log a} = (e^{\log a})^x = a^x$

(ii) Now $\log_a x = y$ if and only if $x = a^y = e^{y \log a}$. But that's equivalent to $\log x = y \log a$. Hence

$$\log_a x = y = \log x / \log a$$

3   (i) The three shaded areas below are clearly in increasing order and the middle one is the area for $\log x$ with the area for $\log x_0$ rubbed out.

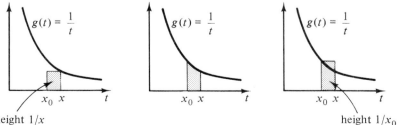

Hence

$$(x - x_0)\frac{1}{x} < \log x - \log x_0 < (x - x_0)\frac{1}{x_0}$$

(and the same inequalities can be deduced if $x < x_0$). Therefore

$$\frac{1}{x} < \frac{\log x - \log x_0}{x - x_0} < \frac{1}{x_0} \quad \text{and} \quad \frac{1}{x_0} < \frac{\log x - \log x_0}{x - x_0} < \frac{1}{x}$$

$$(x > x_0) \qquad\qquad\qquad (x < x_0)$$

So what happens as x tends to x_0? If (x_n) is any sequence ($\neq x_0$) converging to x_0 then

$$-\left|\frac{1}{x_n} - \frac{1}{x_0}\right| < \frac{\log x_n - \log x_0}{x_n - x_0} - \frac{1}{x_0} < \left|\frac{1}{x_n} - \frac{1}{x_0}\right|$$

$$\underbrace{\phantom{-\left|\frac{1}{x_n} - \frac{1}{x_0}\right|}}_{\downarrow} \qquad \underbrace{\phantom{\frac{\log x_n - \log x_0}{x_n - x_0} - \frac{1}{x_0}}}_{\therefore\ \downarrow} \qquad \underbrace{\phantom{\left|\frac{1}{x_n} - \frac{1}{x_0}\right|}}_{\downarrow}$$

$$0 \qquad\qquad 0 \qquad\qquad 0$$

and so

$$\lim_{x \to x_0} \frac{\log x - \log x_0}{x - x_0} = \frac{1}{x_0}$$

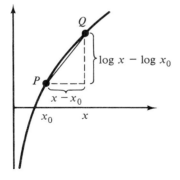

(ii) gradient of $PQ = \dfrac{\log x - \log x_0}{x - x_0}$

and by (i) this tends to $1/x_0$ as x tends to x_0. It therefore seems that the gradient of the log function at $(x_0, \log x_0)$ is $1/x_0$, but we'll be looking at that again in chapter 4.

Page 132

1   To show that the cosine function is continuous and that the given limit is as stated you could go back to basics and imitate our proofs for the sine function. Alternatively you could note that

$$\cos x = \sin(\pi/2 - x)$$

Hence the cosine function is a composition $f \circ g$ of two continuous functions ($f(x) = \sin x$ and $g(x) = \pi/2 - x$) and is itself continuous.

Also

$$\lim_{x \to x_0} \frac{\cos x - \cos x_0}{x - x_0} = \lim_{x \to x_0} \frac{\sin(\pi/2 - x) - \sin(\pi/2 - x_0)}{-((\pi/2 - x) - (\pi/2 - x_0))}$$

and as $x \to x_0$ if and only if $\pi/2 - x \to \pi/2 - x_0$ we can replace $\pi/2 - x$ by y and $\pi/2 - x_0$ by y_0 to deduce that the limit equals

$$\lim_{\pi/2 - x \to \pi/2 - x_0} \frac{\sin(\pi/2 - x) - \sin(\pi/2 - x_0)}{-((\pi/2 - x) - (\pi/2 - x_0))} = \lim_{y \to y_0} -\frac{\sin y - \sin y_0}{y - y_0}$$

$$= -\cos y_0$$
$$= -\cos(\pi/2 - x_0)$$
$$= -\sin x_0$$

as required.

2   Let $\cos^{-1} x = y$ so that in particular $0 \leqslant y \leqslant \pi$. Then

$$x = \cos y = \sin(\pi/2 - y) \quad \text{and} \quad -\pi/2 \leqslant \pi/2 - y \leqslant \pi/2$$

Therefore

$$\sin^{-1} x = \frac{\pi}{2} - y = \frac{\pi}{2} - \cos^{-1} x$$

as required.

3   Simply replace the cosh and sinh by their equivalent expressions involving the exponential to give

$$\cosh^2 x - \sinh^2 x = \left(\frac{e^x + e^{-x}}{2}\right)^2 - \left(\frac{e^x - e^{-x}}{2}\right)^2$$

$$= \tfrac{1}{4}((e^{2x} + 2 + e^{-2x}) - (e^{2x} - 2 + e^{-2x})) = 1$$

and similarly multiplying out and tidying up soon yields.

$$\sinh x \cosh y + \sinh y \cosh x = \frac{e^x - e^{-x}}{2} \frac{e^y + e^{-y}}{2} + \frac{e^y - e^{-y}}{2} \frac{e^x + e^{-x}}{2}$$

$$= \frac{e^{x+y} - e^{x-y}}{2} = \sinh(x + y)$$

In particular $\sinh 2x = 2 \sinh x \cosh x$ and, in a similar fashion, it can be shown that $\cosh 2x = \cosh^2 x + \sinh^2 x$. Therefore

$$\tanh 2x = \frac{\sinh(2x)}{\cosh(2x)} = \frac{2 \sinh x \cosh x}{\cosh^2 x + \sinh^2 x}$$

$$= \frac{2 \dfrac{\sinh x}{\cosh x}}{1 + \dfrac{\sinh^2 x}{\cosh^2 x}} = \frac{2 \tanh x}{1 + \tanh^2 x}$$

4   If $y = \sinh^{-1} x$ then $x = \sinh y = \tfrac{1}{2}(e^y - e^{-y})$. Multiplying through this equation by e^y soon gives the required quadratic in e^y:

$$e^{2y} - 2x\, e^y - 1 = 0$$

with solutions

$$e^y = x + \sqrt{(1 + x^2)} \quad (>0) \quad \text{and} \quad e^y = x - \sqrt{(1 + x^2)} \quad (<0)$$

But as e^y is positive only the former solution is possible and it follows that $y = \log(x + \sqrt{(1 + x^2)})$ as required.

5   The sine function takes the value 1 when the variable is $\pi/2, 5\pi/2, 9\pi/2, \ldots$ and it takes the value -1 at $3\pi/2, 7\pi/2, 11\pi/2, \ldots$. So consider the sequence

$$\frac{1}{\pi}, \quad \frac{2}{3\pi}, \quad \frac{2}{5\pi}, \quad \frac{2}{7\pi}, \quad \frac{2}{9\pi}, \ldots \to 0$$

Then the sequence

$$f\left(\frac{2}{\pi}\right), \quad f\left(\frac{2}{3\pi}\right), \quad f\left(\frac{2}{5\pi}\right), \quad f\left(\frac{2}{7\pi}\right), \quad f\left(\frac{2}{9\pi}\right), \ldots$$
$$\| \qquad\quad \| \qquad\quad \| \qquad\quad \| \qquad\quad \|$$
$$1 \qquad -1 \qquad 1 \qquad -1 \qquad 1$$

does not converge to $f(0)$ (it doesn't even converge) and so f is not continuous at $x = 0$. In fact it is continuous elsewhere because for $x \neq 0$ f is a composition of continuous functions.

In a similar way g is continuous at non-zero points and so it simply remains to prove that g is continuous at $x = 0$. Note that since sine always lies between -1 and 1 it follows that $g(x)$ always lies between $-x$ and x. Hence

$$\text{if} \qquad x_1, x_2, x_3, \ldots \to 0$$
$$\text{then} \quad -x_n \leqslant g(x_n) \leqslant x_n$$
$$\downarrow \qquad\quad \downarrow \qquad\quad \downarrow$$
$$0 \quad \therefore \downarrow \quad 0$$
$$0$$

and

$$\text{then} \quad g(x_1), g(x_2), g(x_3), \ldots \to 0 = g(0)$$

Hence g is continuous at 0. The graphs of f and g are as shown:

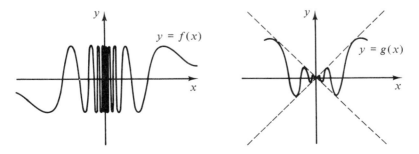

Page 142

1   Assume for example that $a < b$. Then the process defined by the flow-chart comes to a halt when μ happens to hit exactly upon a root or when $b_n - a_n$ is

less than 2ε. Since the sequence

$$b_1 - a_1, b_2 - a_2, b_3 - a_3, b_4 - a_4, \ldots$$

equals

$$b_1 - a_1, \quad \tfrac{1}{2}(b_1 - a_1), \quad \tfrac{1}{4}(b_1 - a_1), \quad \tfrac{1}{8}(b_1 - a_1), \ldots \to 0$$

it follows that one of its terms will eventually be less than 2ε and that the process will stop. At that stage we would have

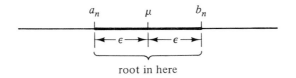

and μ *is* within ε of a root.

Applying the method to the function $f(x) = \tan x - x$ gives, by a few tries on the calculator, $f(4) < 0$ and $f(4.5) > 0$. Hence

$$l_1 = 4 \qquad\qquad m_1 = 4.5$$

$$f(4.25) < 0$$

$$l_2 = 4.25 \qquad\qquad m_2 = 4.5$$

$$f(4.375) < 0$$

$$l_3 = 4.375 \qquad\qquad m_3 = 4.5$$

$$f(4.4375) < 0$$

$$l_4 = 4.4375 \qquad\qquad m_4 = 4.5$$

$$f(4.46875) < 0$$

$$l_5 = 4.46875 \qquad\qquad m_5 = 4.5$$

$$f(4.484375) < 0$$

$$l_6 = 4.484375 \qquad\qquad m_6 = 4.5$$

$$f(4.4921875) < 0$$

$$l_7 = 4.4921875 \qquad\qquad m_7 = 4.5$$

$$f(4.4960938) > 0$$

$$l_8 = 4.492\,187\,5 \qquad m_8 = 4.4960938$$

$$\therefore \quad 4.4941 \text{ is within } 0.002 \text{ of the required root}$$

2   Let $f(x) = ax^3 + bx^2 + cx + d$ where $a > 0$ (the case where $a < 0$ following easily). Then we shall try to show that there exists an x with $f(x) > 0$ and another x with $f(x) < 0$. As you might guess 'for large x' it follows that '$f(x)$ is large'. To show formally that $\lim_{x \to \infty} f(x) = \infty$ note that

if $x_1, x_2, x_3, \ldots \to \infty$

then we may assume that the terms are positive and (by exercise 7 on page 51) we can deduce that

$$\frac{1}{x_1}, \frac{1}{x_2}, \frac{1}{x_3}, \ldots \to 0$$

and hence that eventually the $1/f(x_n)$ are positive and $\to 0$:

$$\frac{1}{f(x_n)} = \frac{1}{ax_n^3 + bx_n^2 + cx_n + d} = \frac{1/x_n^3}{a + b/x_n + c/x_n^2 + d/x_n^3} \to \frac{0}{a + 0 + 0 + 0} = 0$$

Therefore (by that same exercise again) we have

then $f(x_1), f(x_2), f(x_3), \ldots \to \infty$

Hence $\lim_{x \to \infty} f(x) = \infty$ and in particular $f(\beta) > 0$ for some number b.

In a similar way $\lim_{x \to -\infty} f(x) = -\infty$ and so there exists some number α with $f(\alpha) < 0$. But f is continuous and its domain includes $[\alpha, \beta]$ and so by the intermediate value theorem there is some γ between α and β with $f(\gamma) = 0$. Therefore the equation $f(x) = 0$ has a root.

3   Let f be a continuous function with domain $[a, b]$ and with range contained in $[a, b]$. Then if we define a function g by $g(x) = f(x) - x$ for each $x \in [a, b]$ it follows that g is also continuous and that

$$g(a) = \underbrace{f(a)}_{\substack{\in [a, b] \\ \therefore \geqslant a}} - a \geqslant 0 \quad \text{and} \quad g(b) = \underbrace{f(b)}_{\substack{\in [a, b] \\ \therefore \leqslant b}} - b \leqslant 0$$

It follows that $g(x_0) = 0$ for some $x_0 \in [a, b]$ (for either $g(a) = 0$ or $g(b) = 0$ or, failing that, we can apply the intermediate value theorem to g to show that there exists a c between a and b with $g(c) = 0$). But if $g(x_0) = 0$ then $f(x_0) = x_0$ as required.

If $f(x) = \cos x$ $(x \in [0, \pi/2])$ then f's range is $[0, 1]$ which is a subset of $[0, \pi/2]$. So by the previous part there will exist an $x_0 \in [0, \pi/2]$ with $f(x_0) = x_0$. Taking $x_1 = 0$ and applying the technique of exercise 3 on page 107 gives (approximately) the sequence

$$0, 1 \, (= \cos 0), \quad 0.5403 \, (= \cos 1), \quad 0.85755, \ldots$$

and by simply repeatedly pushing the 'cos' button on the calculator we see that the sequence seems to converge to approximately 0.73908 (or 0.739 to three decimal places). In fact $\cos(0.7385) > 0.7385$ and $\cos(0.7395) < 0.7395$ which, if you think about it, confirms that the answer is correct to three decimal places.

The following graphs imply that the root we've just been looking for is the unique $x \in \mathbb{R}$ for which $x = \cos x$.

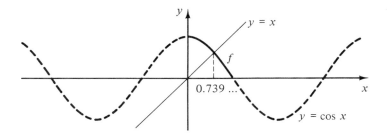

4   We'll do (i) only, as (ii) is very similar.

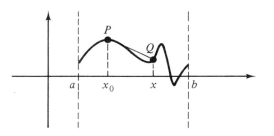

If P and Q are as shown, where the maximum of the function is actually reached at P, then clearly $f(x) \leqslant f(x_0)$ and so

$$\text{gradient of } PQ = \frac{f(x) - f(x_0)}{x - x_0} \leqslant 0$$

as required.

Chapter 4

Page 149

1   Multiplying out the brackets is straightforward and all but two of the terms cancel to give the required expression. Hence

$$\lim_{x \to x_0} \frac{x^m - x_0{}^m}{x - x_0} = \lim_{x \to x_0} (x^{m-1} + x^{m-2}x_0 + x^{m-3}x_0{}^2 + \cdots + x_0{}^{m-1}) = mx_0{}^{m-1}$$
$$\longleftarrow m \text{ terms each tending to } x_0{}^{m-1} \longrightarrow$$

This is precisely the same as saying

$$f'(x_0) = \lim_{x \to x_0} \frac{f(x) - f(x_0)}{x - x_0} = mx_0{}^{m-1}$$

Hence f is differentiable and $f'(x) = mx^{m-1}$ for each x.

2   Given that f is differentiable at x_0 it is clear that

$$g'(x_0) = \lim_{x \to x_0} \frac{g(x) - g(x_0)}{x - x_0} = \lim_{x \to x_0} \frac{kf(x) - kf(x_0)}{x - x_0}$$

$$= k \times \lim \frac{f(x) - f(x_0)}{x - x_0} = k \times f'(x_0)$$

as required.

3   (i) Since f is differentiable at x_0 (and hence continuous there) we have

$$g'(x_0) = \lim_{x \to x_0} \frac{g(x) - g(x_0)}{x - x_0} = \lim_{x \to x_0} \frac{1/f(x) - 1/f(x_0)}{x - x_0}$$

$$= \lim_{x \to x_0} - \frac{1}{f(x)f(x_0)} \frac{f(x) - f(x_0)}{x - x_0} = - \frac{1}{(f(x_0))^2} f'(x_0)$$

(ii) Applying (i) to $f(x) = x^m$ and using exercise 1 we get $g(x) = 1/f(x)$ and

$$g'(x) = - \frac{f'(x)}{(f(x))^2} = - \frac{mx^{m-1}}{(x^m)^2} = -mx^{-m-1}$$

So $f(x) = x^n$ has derivative $f'(x) = nx^{n-1}$ for:

(a) positive integers n (exercise 1);

(b) negative integers n (above); and

(c) $n = 0$ (since then f is a constant, namely 1, and it's easy to check that $f'(x) = 0$).

4   (i) It is clear from the picture below that the reflection of a line of gradient α ($\neq 0$) in the line $y = x$ has gradient $1/\alpha$. Hence (looking at the figure in the exercise) if $g'(f(x_0)) = \alpha$ then we'd expect $f'(x_0)$ to be $1/\alpha$.

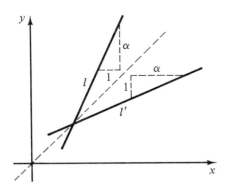

Formally,

if $x_1, x_2, x_3, \ldots \to x_0$

in f's domain, $\neq x_0$

then (since f is continuous and one-to-one with inverse g)

$$\underbrace{f(x_1), f(x_2), f(x_3), \ldots}_{\text{in } g\text{'s domain, } \neq f(x_0)} \to f(x_0)$$

and so by the differentiability of g at $f(x_0)$

$$\frac{g(f(x_n)) - g(f(x_0))}{f(x_n) - f(x_0)} \to g'(f(x_0)) \quad \text{as } n \to \infty$$

i.e.

$$\frac{x_n - x_0}{f(x_n) - f(x_0)} \to \alpha \quad \text{as } n \to \infty$$

But

then $\quad \dfrac{f(x_n) - f(x_0)}{x_n - x_0} \to \dfrac{1}{\alpha} \quad \text{as } n \to \infty$

Hence from the definition of the derivative we have shown that

$$f'(x_0) = \lim_{x \to x_0} \frac{f(x) - f(x_0)}{x - x_0} = \frac{1}{\alpha}$$

as required.

(ii) Apply (i) to the functions given by

$$f(x) = x^{1/m} \quad x > 0 \quad \text{and} \quad g(x) = x^m \quad x > 0$$

to give

$$f'(x) = \frac{1}{g'(f(x))} = \frac{1}{m(x^{1/m})^{m-1}} = \frac{1}{mx^{(m-1)/m}} = \frac{1}{m} x^{(1/m)-1}$$

as required.

Page 157

1   (i) $f(x) = a^x = e^{x \log a} \quad \Rightarrow \quad f'(x) = \log a \times e^{x \log a} = \log a \times a^x$

$$g(h(x)) \qquad\qquad h'(x) \times g'(h(x))$$

(ii) $g(x) = \log_a x = \dfrac{\log x}{\log a} = \dfrac{1}{\log a} \times \log x \quad \Rightarrow \quad g'(x) = \dfrac{1}{\log a} \times \dfrac{1}{x} = \dfrac{1}{x \log a}$

constant

2   It is easy to check that

$$h(x) = \cosh x = \tfrac{1}{2}(e^x + e^{-x})$$
$$\Rightarrow \quad h'(x) = \tfrac{1}{2}(e^x - e^{-x}) = \sinh x = \sqrt{(\cosh^2 x - 1)}$$

Hence if $j(x) = \cosh^{-1} x$ then by the rule for differentiating inverse functions (exercise 4 on page 150) we see that

$$h'(j(x)) = \sqrt{(x^2 - 1)} \quad \text{and} \quad j'(x) = \frac{1}{\sqrt{(x^2 - 1)}}$$

Hence for the given composite function $f(x) = \cosh^{-1}(x/2)$ we see that

$$f'(x) = \tfrac{1}{2} \frac{1}{\sqrt{((x/2)^2 - 1)}} = \frac{1}{\sqrt{(x^2 - 4)}} \qquad x > 2$$

Now $g(x) = \log(x + \sqrt{(x^2 - 4)})$ is also a composite function and using the chain rule twice gives

$$g'(x) = \frac{1}{x + (x^2 - 4)^{1/2}} \times (1 + x(x^2 - 4)^{-1/2})$$

$$= \frac{1}{x + \sqrt{(x^2 - 4)}} \times \frac{x + \sqrt{(x^2 - 4)}}{\sqrt{(x^2 - 4)}} = \frac{1}{\sqrt{(x^2 - 4)}} \qquad x > 2$$

Hence $f' = g'$ as claimed.

That only says that the functions f and g have the same gradient at each $x > 2$, and rough sketches of their graphs are illustrated.

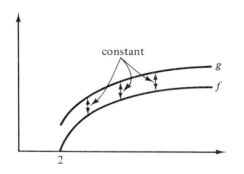

We shall see in the next exercises that in fact f and g must differ by a constant amount.

3   (i) $F'(x) = -2x \times \tfrac{1}{2}(1 - x^2)^{-1/2} = -\dfrac{x}{\sqrt{(1 - x^2)}} = \dfrac{\cos t}{-\sin t} = \dfrac{g'(t)}{f'(t)}$

as claimed.

(ii) In general if $y = g(t)$ and $y = F(x)$ where $x = f(t)$ then $y = g(t) = F(f(t))$. Hence by the chain rule

$$g'(t) = f'(t) \times F'(f(t))$$

and so

$$F'(x) = F'(f(t)) = \frac{g'(t)}{f'(t)}$$

as claimed.

In the alternative notation this says that

$$\frac{dy}{dx} = \frac{dy/dt}{dx/dt}$$

which again looks like a piece of correct algebra.

4   For $x \neq 0$ for all three of the functions f, g and h are simple combinations of differentiable functions and are clearly differentiable. However at $x = 0$ the function f is not continuous (exercise 5 on page 132) and so it is certainly not differentiable there. To consider the differentiability of g and h at 0 note that

$$\lim_{x \to 0} \frac{g(x) - g(0)}{x - 0} = \lim_{x \to 0} \frac{x \sin(1/x)}{x} = \lim_{x \to 0} \sin(1/x) = \lim_{x \to 0} f(x)$$

and, as we saw in the solution of exercise 5 on page 259, that limit does not exist. Hence g is not differentiable at 0. On the other hand

$$\lim_{x \to 0} \frac{h(x) - h(0)}{x - 0} = \lim_{x \to 0} \frac{x^2 \sin(1/x)}{x} = \lim_{x \to 0} x \sin(1/x)$$

$$= \lim_{x \to 0} g(x) = 0$$

(since, by that same previous exercise, g is continuous at 0). Thus h is differentiable at $x = 0$ and $h'(0) = 0$, as illustrated.

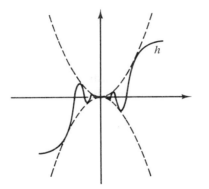

5        $$\lim_{x \to x_0} \frac{f(x)}{g(x)} = \lim_{x \to x_0} \frac{\overset{0}{\overset{\|}{f(x)}} - f(x_0)}{g(x) - g(x_0)} = \lim_{x \to x_0} \frac{\dfrac{f(x) - f(x_0)}{x - x_0}}{\underset{0}{\underset{\|}{g(x)}} - g(x_0)}$$

$$= \frac{\displaystyle\lim_{x \to x_0} \frac{f(x) - f(x_0)}{x - x_0}}{\displaystyle\lim_{x \to x_0} \frac{g(x) - g(x_0)}{x - x_0}} = \frac{f'(x_0)}{g'(x_0)}$$

(i) This example does not require the result established above because the limit of the denominator is not 0:

$$\lim_{x \to 1} \frac{e^x}{x^2 + 1} = \frac{\displaystyle\lim_{x \to 1} e^x}{\displaystyle\lim_{x \to 1} (x^2 + 1)} = \frac{e}{2}$$

(ii) Here the limit of the numerator and the denominator are both 0 and so we differentiate the top and bottom and apply the above result to give

$$\lim_{x \to 0} \frac{\sin x}{e^x - 1} = \frac{\cos 0}{e^0} = 1$$

Page 165

1   (i)

$f(x) = x^4$	$g(x) = \sin x$
$f'(x) = 4x^3$	$g'(x) = \cos x$
$f''(x) = 12x^2$	$g''(x) = -\sin x$
$f^{(3)}(x) = 24x$	$g^{(3)}(x) = -\cos x$
$f^{(4)}(x) = 24$	$g^{(4)}(x) = \sin x$
$f^{(5)}(x) = 0$	$g^{(5)}(x) = \cos x$
$f^{(6)}(x) = 0$	$g^{(6)}(x) = -\sin x$
$\vdots$	$\vdots$
(all zero hereafter)	(pattern repeats every four)

(ii) $f(x) = x^4 - 2x^2$ and so

$$f'(x) = 4x^3 - 4x = 4x(x^2 - 1)$$

which is zero when x is $-1, 0$ or 1. It is then easy to construct a sketch graph of f, as illustrated below. Note that $f''(x) = 12x^2 - 4$ which is positive at $x = \pm 1$ (the local minima) and negative at $x = 0$ (the local maximum).

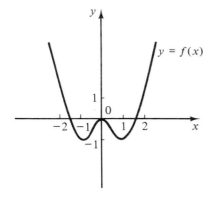

(iii) $f'(x) = 3x^2$ and so $f'(0) = 0$. Also, as in the solution to exercise 4 above, $h'(0) = 0$. So f and h both have a stationary point at $x = 0$. The well-known graph of f and the sketch graph of h on the previous page show that neither of the functions has a local maximum or minimum at $x = 0$. In fact both the functions are *odd* which means for example that $f(-x) = -f(x)$. So if f (or h) is positive for some x then it is negative for $-x$. Such functions clearly have $f(0) = 0$ and they cannot have a local maximum or minimum at $x = 0$.

In the case of f we have that $f''(x) = 6x$ which actually changes sign at $x = 0$ (i.e. the graph of f'' crosses the x-axis). So f has a 'point of inflection' (or 'kink') there.

Note for future reference that the function g given by $g(x) = x^4$ has $g'(x) = 4x^3$ and $g''(x) = 12x^2$ and so $g'(0) = g''(0) = 0$. Hence g has a stationary point at $x = 0$ and, despite the fact that $g''(0) = 0$, g has a local minimum (rather than a point of inflection) at $x = 0$. So if you think that g'' being zero at a stationary point means that the function has an inflection there then kindly **un**think it now. The correct details will be found in the next section.

2   The proof of Leibniz' rule is by induction on n, the case $n = 1$ being

$$(fg)' = \binom{1}{0}fg' + \binom{1}{1}f'g = fg' + f'g$$

which is the usual rule for differentiating a product, as established on page 152.

The typical next step in a proof by induction is to assume that $k > 1$ and that the result is known for $n = k - 1$ and then to deduce the result for $n = k$. Keen readers will be able to do this but if we try to present all the details here they will go off the side of the page, so we just illustrate how the result for $n = 3$ implies the result for $n = 4$: the ideas easily extend to any k.

Assume then that

$$(fg)^{(3)} = \binom{3}{0}fg^{(3)} + \binom{3}{1}f'g^{(2)} + \binom{3}{2}f^{(2)}g' + \binom{3}{3}f^{(3)}g$$

Then $fg^{(3)}$ is the sum of four differentiable products and so it is differentiable. Thus fg is differentiable four times and its derivative is obtained by differentiating the above to give

$$(fg)^{(4)} = \binom{3}{0}[fg^{(4)} + f'g^{(3)}] + \binom{3}{1}[f'g^{(3)} + f^{(2)}g^{(2)}]$$

$$+ \binom{3}{2}[f^{(2)}g^{(2)} + f^{(3)}g'] + \binom{3}{3}[f^{(3)}g' + f^{(4)}g]$$

$$= \binom{3}{0}fg^{(4)} + \left[\binom{3}{0} + \binom{3}{1}\right]f'g^{(3)} + \left[\binom{3}{1} + \binom{3}{2}\right]f^{(2)}g^{(2)}$$

$$+ \left[\binom{3}{2} + \binom{3}{3}\right]f^{(3)}g' + \binom{3}{3}f^{(4)}g$$

$$= \binom{4}{0}fg^{(4)} + \binom{4}{1}f'g^{(3)} + \binom{4}{2}f^{(2)}g^{(2)}$$

$$+ \binom{4}{3}f^{(3)}g' + \binom{4}{4}f^{(4)}g$$

where we have used the facts (in the case $k = 4$) that

$$\binom{k-1}{0} = \binom{k}{0} (=1) \quad \binom{k-1}{k-1} = \binom{k}{k} (=1) \quad \text{and}$$

$$\binom{k-1}{r-1} + \binom{k-1}{r} = \binom{k}{r}$$

as we saw in the solution of exercise 6 on page 229.

The interested reader can now adapt the above argument to the general inductive step. The key fact is that in differentiating the expression for $(fg)^{(k-1)}$ you generally get terms in $f^{(r)}g^{(k-r)}$ occurring twice giving the total of them as

$$\binom{k-1}{r-1} + \binom{k-1}{r} \quad \left(=\binom{k}{r}\right)$$

In this way it is straightforward to establish Leibniz' rule by induction.

Applying the rule to $f(x) = x^2$ and $g(x) = e^{2x}$ gives $g^{(n)}(x) = 2^n e^{2x}$ and $f^{(3)}(x) = f^{(4)}(x) = \cdots = 0$ and so

$$(fg)^{(7)}(x) = \binom{7}{0}x^2 \times 2^7 e^{2x} + \binom{7}{1}2x \times 2^6 e^{2x} + \binom{7}{2}2 \times 2^5 e^{2x}$$

$$= 64(2x^2 + 14x + 21)\,e^{2x}$$

3   If $f(x) = \sin^{-1}x$ then the usual rule for differentiating an inverse function soon leads to $f'(x) = 1/\sqrt{(1-x^2)}$. Hence applying the mean value theorem to f between 0 and x ($\in (0, 1)$) gives

$$\frac{\sin^{-1}x}{x} = \frac{\sin^{-1}x - \sin^{-1}0}{x - 0} = \frac{f(x) - f(0)}{x - 0} = f'(c) = \frac{1}{\sqrt{(1-c^2)}}$$

for some c between 0 and x. But then $c \in (0, x)$ and so $1/\sqrt{(1-c^2)}$ lies between 1 and $1/\sqrt{(1-x^2)}$. Hence

$$1 \leqslant \frac{\sin^{-1}x}{x} \leqslant \frac{1}{\sqrt{(1-x^2)}} \quad \text{and} \quad x \leqslant \sin^{-1}x \leqslant \frac{x}{\sqrt{(1-x^2)}}$$

as required.

4   By the mean value theorem applied separately to f and g there exist a c and d in (a, b) with

$$f'(c) = \frac{f(b) - f(a)}{b - a} \quad \text{and} \quad g'(d) = \frac{g(b) - g(a)}{b - a}$$

but we certainly cannot be sure that $c = d$, as the examples illustrated below show. There $f(x) = x^2$ and $g(x) = x^3$ ($x \in [0, 1]$) and for f the mean value actually occurs at $c = \frac{1}{2}$ whereas for g the mean value actually occurs at $d = 1/\sqrt{3}$.

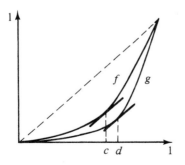

For a valid proof we'll apply Rolle's theorem to the function h given by

$$h(x) = f(x) - \frac{f(b) - f(a)}{g(b) - g(a)}\,g(x) \qquad x \in [a, b]$$

Clearly h is continuous on $[a, b]$ and differentiable on (a, b) since both f and g are. Also

$$h(b) - h(a) = \left(f(b) - \frac{f(b) - f(a)}{g(b) - g(a)} g(b) \right) - \left(f(a) - \frac{f(b) - f(a)}{g(b) - g(a)} g(a) \right)$$

$$= (f(b) - f(a)) - \frac{f(b) - f(a)}{g(b) - g(a)} (g(b) - g(a)) = 0$$

and so $h(a) = h(b)$. Thus h has all the properties necessary for us to be able to apply Rolle's theorem (page 162) to deduce that $h'(c) = 0$ for some $c \in (a, b)$; i.e.

$$h'(c) = f'(c) - \frac{f(b) - f(a)}{g(b) - g(a)} g'(c) = 0 \quad \text{and} \quad \frac{f'(c)}{g'(c)} = \frac{f(b) - f(a)}{g(b) - g(a)}$$

as required.

In the case of the f and g illustrated the value of c can be calculated as $\frac{2}{3}$.

5　　　**If**　$\underbrace{x_1, x_2, x_3, \ldots}_{\neq x_0} \to x_0$

then by exercise 4 above

$$\frac{f(x_n)}{g(x_n)} = \frac{f(x_n) - \overset{\overset{0}{\|}}{f(x_0)}}{g(x_n) - \underset{\underset{0}{\|}}{g(x_0)}} = \frac{f'(c_n)}{g'(c_n)}$$

for some c_n between x_0 and x_n. But as $\underbrace{x_1, x_2, x_3, \ldots}_{\neq x_0} \to x_0$ then so does the

sandwiched sequence $\underbrace{c_1, c_2, c_3, \ldots,}_{\neq x_0} \to x_0$. Hence, as $\lim(f'(x)/g'(x))$ exists it

follows that

$$\textbf{then} \quad \frac{f(x_n)}{g(x_n)} = \frac{f'(c_n)}{g'(c_n)} \to \lim_{x \to x_0} \frac{f'(x)}{g'(x)}$$

Therefore

$$\lim_{x \to x_0} \frac{f(x)}{g(x)} = \lim_{x \to x_0} \frac{f'(x)}{g'(x)}$$

as required: this is called L'Hopital's rule.

(i) For this part we can either use our new result or the simpler version from exercise 5 on page 157:

$$\lim_{x \to 1} \frac{\log x}{x - 1} = \lim_{x \to 1} \frac{1/x}{1} = 1$$

$$\underbrace{\qquad\qquad\qquad}_{\text{if it exists – which it does!}}$$

(ii) Here we apply L'Hopital's rule twice to give

$$\lim_{x \to 0} \frac{\cos x - 1}{x^2} = \lim_{x \to 0} \frac{-\sin x}{2x} = \lim_{x \to 0} \frac{-\cos x}{2} = -\tfrac{1}{2}$$

You might like to check the plausibility of this answer by calculating $(\cos x - 1)/x^2$ for some small x (in radians of course). For example when $x = 0.01$ the expression is approximately $-0.499\,995$.

(iii) $\displaystyle \lim_{x \to 0} \frac{\log(1 + \alpha x)}{x} = \lim_{x \to 0} \frac{\alpha/(1 + \alpha x)}{1} = \alpha$

(iv) Since $(\log(1 + \alpha x))/x \to \alpha$ as $x \to 0$ we can use the continuity of the exponential function to deduce that

$$(1 + \alpha x)^{1/x} = e^{(\log(1 + \alpha x))/x} \to e^{\alpha} \quad \text{as} \quad x \to 0$$

In particular

$$(1 + \alpha/n)^n \to e^{\alpha} \quad \text{as} \quad n \to \infty$$

(We suspected much earlier – in exercise 5 on pages 37 and 38 – that $(1 + 1/n)^n \to e$ as $n \to \infty$ and now at last we have proved that and much more.)

6   (i) Let f be a differentiable function with $f'(x) > 0$ for each x. Then given any a and b in f's domain with $a < b$ apply the mean value theorem to f on $[a, b]$ to deduce that for some $c \in (a, b)$

$$\frac{f(b) - f(a)}{b - a} = f'(c) \quad \text{and} \quad f(b) - f(a) = \underbrace{f'(c)}_{>0}\underbrace{(b - a)}_{>0}$$

Hence if $a < b$ (in f's domain) then $f(a) < f(b)$; i.e. f is strictly increasing. A similar argument works for (ii), (iii) and (iv).

Now if $f(x) = x - \sin x$ $(x \in \mathbb{R})$ then

$$f'(x) = 1 - \cos x \geqslant 0 \quad (x \in \mathbb{R})$$

and so by (ii) f is increasing. In particular

$$0 \leqslant x \quad \Rightarrow \quad f(0) \leqslant f(x) \quad \Rightarrow \quad 0 \leqslant x - \sin x \quad \Rightarrow \quad \sin x \leqslant x$$

as required.

7   (i) Clearly the function $f - g$ has zero derivative throughout its domain (which is an interval) and so by the corollary on page 165 $f - g$ is a constant function. Hence there is a number k for which $f(x) - g(x) = k$ (and $f(x) = g(x) + k$) for each x.

(ii) As we saw in exercise 2 on page 157 the functions given by

$$f(x) = \cosh^{-1}(x/2) \quad x \geqslant 2 \quad \text{and} \quad g(x) = \log(x + \sqrt{(x^2 - 4)}) \quad x \geqslant 2$$

have the same derivative for each $x > 2$. Hence by (i) there is some number k with

$$f(x) = g(x) + k \quad \text{for each } x > 2$$

Since f and g are continuous at $x = 2$ it follows also that

$$0 = f(2) = \lim_{x \to 2} f(x) = \lim_{x \to 2} (g(x) + k) = g(2) + k = \log 2 + k$$

and so $k = -\log 2$. Therefore

$$\cosh^{-1}(x/2) = \log(x + \sqrt{(x^2 - 4)}) - \log 2 \quad \text{for each } x \geqslant 2$$

as required.

8   By exercise 6 (i) above since $g'(x) = g(x) > 0$ for each x we can deduce that g is strictly increasing and hence that it has an inverse function F say. Now we know how to differentiate inverse functions from exercise 4 on page 150 and we deduce that

$$F'(x) = \frac{1}{g'(F(x))} = \frac{1}{g(F(x))} = \frac{1}{x} = f'(x) \quad \text{for each } x > 0$$

Hence by exercise 7(i) f and F $(=g^{-1})$ differ by a constant.

(i) In particular we know from pages 118 and 110 that the functions given by

$$f_1(x) = \log x \quad x > 0 \quad \text{and} \quad g_1(x) = \exp x \quad x \in \mathbb{R}$$

satisfy

$$f_1'(x) = 1/x \quad x > 0 \quad \text{and} \quad g_1'(x) = g_1(x) \quad x \in \mathbb{R}$$

and so the log function f_1 and the inverse g_1^{-1} of the exponential function differ by a constant. However we also know that $f_1(1) = \log 1 = 0$ and $g_1(0) = \exp 0 = 1$. Hence $g_1^{-1}(1) = 0$ and it follows that for each $x > 0$

$$f_1(x) - g_1^{-1}(x) = \text{constant} = f_1(1) - g_1^{-1}(1) = 0$$

Therefore $f_1 = g_1^{-1}$; i.e. the log function is indeed the inverse of the exponential function.

(ii) Similarly f_1 $(=\log)$ and g^{-1} differ by a constant; i.e. there exists a number k with

$$\log x = g^{-1}(x) + k \qquad x > 0$$

Hence

$$g(x) = e^{x+k}$$

as required.

(As usual one way of working out the inverse of g^{-1} is by use of a diagram:

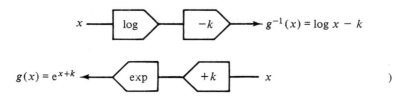

)

Page 178

1   Let $f(x) = \cos x$ and $g(x) = \cosh x$ $(x \in \mathbb{R})$. Then both these functions can be differentiated arbitrarily often to give

$$f'(x) = -\sin x, \quad f''(x) = -\cos x, \quad f^{(3)}(x) = \sin x, \quad f^{(4)}(x) = \cos x, \quad \ldots$$
$$f'(0) = 0 \qquad\quad f''(0) = -1, \qquad\quad f^{(3)}(0) = 0, \qquad\quad f^{(4)}(0) = 1, \qquad \ldots$$

and

$$g'(x) = \sinh x, \quad g''(x) = \cosh x, \quad g^{(3)}(x) = \sinh x, \quad g^{(4)}(x) = \cosh x, \ldots$$
$$g'(0) = 0, \qquad\quad g''(0) = 1, \qquad\quad g^{(3)}(0) = 0, \qquad\quad g^{(4)}(0) = 1, \qquad \ldots$$

and so their Taylor series are

$$1 - \frac{x^2}{2!} + \frac{x^4}{4!} - \frac{x^6}{6!} + \cdots \quad \text{and} \quad 1 + \frac{x^2}{2!} + \frac{x^4}{4!} + \frac{x^6}{6!} + \cdots$$

(and you will notice their similarity).

By the theorem the difference between

$$\cos x_0 \quad \text{and} \quad 1 - \frac{x_0^2}{2!} + \frac{x_0^4}{4!} - \cdots \pm \frac{x_0^n}{n!} \quad \text{is} \quad \frac{x_0^{n+1}}{(n+1)!} f^{(n+1)}(c) = \pm \frac{x_0^{n+1}}{(n+1)!} \sin c$$

for some c between 0 and x_0. As in the worked examples for $\sin x$ on page 175 this difference has modulus less than or equal to $k^{n+1}/(n+1)!$ for any x_0 in the interval $[-k, k]$. But $k^{n+1}/(n+1)!$ tends to 0 as $n \to \infty$ and so certainly there exists an even integer n with $k^{n+1}/(n+1)! < \varepsilon$. In that case the nth degree polynomial approximation for $\cos x$ differs from $\cos x$ by less than ε throughout $[-k, k]$.

Similarly the difference between

$$\cosh x_0 \quad \text{and} \quad 1 + \frac{x_0^2}{2!} + \frac{x_0^4}{4!} + \cdots + \frac{x_0^n}{n!} \quad \text{is}$$

$$\frac{x_0^{n+1}}{(n+1)!} g^{(n+1)}(c) = \frac{x_0^{n+1}}{(n+1)!} \sinh x$$

for some c between 0 and x_0. This difference has modulus less than or equal to $k^{n+1} \sinh k/(n+1)!$ for any x_0 in the interval $[-k, k]$, and again this tends to 0 as $n \to \infty$.

2   Here $f(x) = \log(1 + x)$ $(x > -1)$ and so $f(0) = 0$ and

$$f'(x) = (1 + x)^{-1}, \; f''(x) = -(1 + x)^{-2}, \; f^{(3)}(x) = 2(1 + x)^{-3}, \; f^{(4)} = -3!(1 + x)^{-4}, \ldots$$
$$f'(0) = 1, \qquad\quad f''(0) = -1, \qquad\quad f^{(3)}(0) = 2, \qquad\quad f^{(4)} = -3! \qquad\quad \ldots$$

and therefore the coefficient of x^n in the Taylor series of f is $(-1)^{n+1}(n-1)!/n!$ which is simply $(-1)^{n+1}/n$. Hence the Taylor series (about $x = 0$) is

$$x - \frac{x^2}{2} + \frac{x^3}{3} - \frac{x^4}{4} + \cdots$$

The nth degree polynomial differs from $\log(1 + x_0)$ by

$$\frac{x_0^{n+1}}{(n+1)!} f^{(n+1)}(c) = \pm \frac{x_0^{n+1}}{(n+1)(1 + c)^{n+1}} \quad \text{for some } c \text{ between 0 and } x_0$$

For $x_0 \in [0, 1]$ For $x_0 \in [-\frac{1}{2}, 0)$

$$\frac{\overbrace{|x_0|^{n+1}}^{\leqslant 1}}{\underbrace{(n+1)(1 + c)^{n+1}}_{\geqslant 1}} \qquad\qquad\qquad \frac{\overbrace{|x_0|^{n+1}}^{\leqslant \frac{1}{2}}}{\underbrace{(n+1)(1 + c)^{n+1}}_{\geqslant \frac{1}{2}}}$$

so in each case the modulus of that difference is at most $1/(n + 1)$. Since that tends to 0 as $n \to \infty$ we can certainly find an n with $1/(n + 1) < \varepsilon$. Then the nth degree polynomial approximation will differ from $\log(1 + x)$ by less than ε throughout $[-\frac{1}{2}, 1]$.

(We shall learn in the next chapter that the interval $[-\frac{1}{2}, 1]$ can be extended to $(-1, 1]$.)

3　　　　(Taylor series are usually studied about an arbitrary 'centre' $x = a$ but the point of this exercise is to show that studying Taylor series 'about $x = 0$' is quite sufficient. The Taylor series about $x = 0$ is sometimes known as the *Maclaurin expansion*.)

By the theorem (applied to f and $x_0 - a$ rather than x_0) $f(x_0 - a)$ and $f(0) + f'(0)(x_0 - a) + (f''(0)/2!)(x_0 - a)^2 + \cdots + (f^{(n)}(0)/n!)(x_0 - a)^n$ differ by $(x_0 - a)^{n+1} f^{(n+1)}(d)/(n + 1)!$ for some d between 0 and $x_0 - a$. But since $f(x) = g(x + a)$ it follows that $f'(0) = g'(a)$ etc and

$$f^{(n+1)}(d) = g^{(n+1)}(d + a) = g^{(n+1)}(c)$$

where $c = d + a$ is between a and x_0. Therefore

$$f(x_0 - a) = f(0) + f'(0)(x_0 - a) + \frac{f''(0)}{2!}(x_0 - a)^2 + \cdots$$
$$+ \frac{f^{(n)}(0)}{n!}(x_0 - a)^n + \frac{(x_0 - a)^{n+1}}{(n+1)!} f^{(n+1)}(d)$$

easily translates to

$$g(x_0) = g(a) + g'(a)(x_0 - a) + \frac{g''(0)}{2!}(x_0 - a)^2 + \cdots$$
$$+ \frac{g^{(n)}(a)}{n!}(x_0 - a)^n + \frac{(x_0 - a)^{n+1}}{(n+1)!} g^{(n+1)}(c)$$

as required.

The Taylor series for the log function about $x = 1$ is easily found to be

$$(x - 1) - \frac{(x - 1)^2}{2} + \frac{(x - 1)^3}{3} - \frac{(x - 1)^4}{4} + \cdots$$

which can be compared with the series about $x = 0$ of the function $\log(1 + x)$ in the previous exercise.

4　　　　Now that we have established the general form of Taylor's theorem (previous exercise) the proof of the last theorem concerning stationary points at $x = 0$ easily translates to the general case $x = a$ giving

$$f(x) - f(a) = \underbrace{\frac{f^{(m)}(c)}{m!}}_{\substack{\text{same sign in} \\ \text{some interval} \\ \text{around } x = a}} (x - a)^m$$

From this it is easy to deduce the required conclusions.

If $f(x) = x^3 e^x$ then

$$f'(x) = (x^3 + 3x^2) e^x \qquad f''(x) = (x^3 + 6x^2 + 6x) e^x$$
$$f^{(3)}(x) = (x^3 + 9x^2 + 18x + 6) e^x$$

So f has stationary points when $(x^3 + 3x^2)\,e^x = 0$; i.e. when $x = 0$ and when $x = -3$.

At $x = 0$ the first non-zero derivative is the third (so it's neither a local maximum nor minimum) and at $x = -3$ the first non-zero derivative is the second (with $f''(-3) = 9\,e^{-3} > 0$, giving a local minimum there).

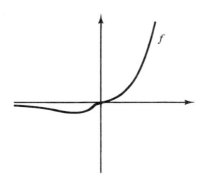

Page 187

1   We use the ratio test (page 69) applied to the series with moduli around each term and then use the fact (page 70) that absolute convergence implies convergence:

(i) $|(n + 1)\text{st term} \div n\text{th term}| = \left|\left(\dfrac{x^{2n}}{n!}\right) \div \left(\dfrac{x^{2n-2}}{(n-1)!}\right)\right| = \dfrac{x^2}{n} \to 0$ as $n \to \infty$

Hence this series is (absolutely) convergent for each x and has infinite radius of convergence. In fact it is clearly the series for e^{x^2}.

(ii) $|(n + 1)\text{st term} \div n\text{th term}| = |(n + 1)x^{n+1} \div nx^n|$

$$= \left(\frac{(n+1)}{n}\right)|x| \to |x| \text{ as } n \to \infty$$

It follows with a little thought that the series converges for $|x| < 1$ and diverges for $|x| \geqslant 1$; hence this series has radius of convergence equal to 1. In fact the series is x multiplied by the derived series of $1 + x + x^2 + \cdots$ (which is the series of $(1 - x)^{-1}$). Hence the given series is the Taylor series of the function f given by $f(x) = x/(1 - x)^2$.

(iii) $|(n + 1)\text{st term} \div n\text{th term}| = n|x| \to \infty$ if $x \neq 0$ and so the series only converges for $x = 0$; i.e. it has radius of convergence equal to 0.

2   This is one of those places where the alternative notation is a little easier to follow: if $y = f(x) = \tan x$ then

$$\frac{dy}{dx} = f'(x) = \sec^2 x = \frac{1}{\cos^2 x} = \frac{\cos^2 x + \sin^2 x}{\cos^2 x} = 1 + \tan^2 x$$

$$= 1 + (f(x))^2 = 1 + y^2$$

Using the chain rule to differentiate this again, and then repeatedly replacing dy/dx

by $1 + y^2$ gives

$$f''(x) = \frac{d^2y}{dx^2} = \frac{dy}{dx} \times 2y = (1 + y^2) \times 2y = 2y + 2y^3$$

$$f^{(3)}(x) = \frac{d^3y}{dx^3} = (1 + y^2)(2 + 6y^2) = 2 + 8y^2 + 6y^4 \quad \text{etc}$$

and you can eventually see that at $x = 0$ we have $y = f(0) = 0$ and $f'(0) = 1$, $f''(0) = 0$, $f^{(3)}(0) = 2$, $f^{(4)}(0) = 0$, $f^{(5)}(0) = 16$ etc. Hence the Taylor series of $\tan x$ starts

$$x + \frac{x^3}{3} + \frac{2x^5}{15} + \cdots$$

More straightforwardly, if $g(x) = \tan^{-1} x$ then $g'(x) = 1/(1 + x^2)$, $g''(x) = -2x/(1 + x^2)^2$ and $g^{(3)}(x) = (6x^2 - 2)/(1 + x^2)^3$, etc, giving the series

$$x - \frac{x^3}{3} + \frac{x^5}{5} - \cdots$$

(i) The product of the series of $\tan x$ and the series of $\cos x$ (assuming we *can* multiply absolutely convergent series in a natural way) is

$$\left(x + \frac{x^3}{3} + \frac{2x^5}{15} + \cdots\right)\left(1 - \frac{x^2}{2!} + \frac{x^4}{4!} - \cdots\right) = x - \frac{x^3}{3!} + \frac{x^5}{5!}\cdots$$

which does seem to be giving the series of $\sin x$.

(ii) Assuming that we can work out compositions of series in a natural way the series of $f(g(x))$ is

$$\left(x - \frac{x^3}{3} + \frac{x^3}{5} - \cdots\right) + \tfrac{1}{3}\left(x - \frac{x^3}{3} + \frac{x^3}{5} - \cdots\right)^3 + \tfrac{2}{15}\left(x - \frac{x^3}{3} + \frac{x^5}{5} - \cdots\right)^5 + \cdots$$

$$= x + 0x^2 + 0x^3 + 0x^4 + 0x^5 + \cdots$$

which seems to simplify to x (which is what you'd expect for these inverse functions). Similarly the series of $g(f(x))$ is

$$\left(x + \frac{x^3}{3} + \frac{2x^5}{15} + \cdots\right) - \tfrac{1}{3}\left(x + \frac{x^3}{3} + \frac{2x^5}{15} + \cdots\right)^3 + \tfrac{1}{5}\left(x + \frac{x^3}{3} + \frac{2x^5}{15} + \cdots\right)^5 - \cdots$$

$$= x + 0x^2 + 0x^3 + 0x^4 + 0x^5 + \cdots$$

which again seems to simplify to x alone.

3   (i) By differentiation of f's series in exercise 2 above we obtain the first few terms of the series for $\sec^2 x$ as

$$1 + x^2 + \frac{2x^4}{3} + \cdots$$

(ii) Similarly differentiating g's series above gives

$$1 - x^2 + x^4 - \cdots$$

which, as expected, is the start of the Taylor series of $(1 + x^2)^{-1}$.

Chapter 5

Page 204

1   (i) We are given that $t_r = \frac{1}{2}r(r+1)$ and so

$$t_r^2 - t_{r-1}^2 = (\tfrac{1}{2}r(r+1))^2 - (\tfrac{1}{2}(r-1)r)^2$$
$$= \tfrac{1}{4}r^2((r+1)^2 - (r-1)^2) = \tfrac{1}{4}r^2 \times 4r = r^3$$

Hence

$$1^3 + 2^3 + \cdots + n^3 = (\overbrace{t_1^2 - t_0^2}^{0}) +$$
$$(t_2^2 - t_1^2) +$$
$$(t_3^2 - t_2^2) +$$
$$\vdots$$
$$(t_n^2 - t_{n-1}^2)$$
$$= t_n^2 = (\tfrac{1}{2}n(n+1))^2$$

as required.

(ii) The usual evenly-spaced partition $(0, 1/n, 2/n, \ldots, (n-1)/n, 1)$ of $[0, 1]$ gives (using the formula for the sum of cubes from above)

$$l_n = \left(\frac{1}{n} \times 0^3\right) + \left(\frac{1}{n} \times \left(\frac{1}{n}\right)^3\right) + \cdots + \left(\frac{1}{n} \times \left(\frac{n-1}{n}\right)^3\right)$$
$$= \frac{1}{n^4}(0^3 + 1^3 + 2^3 + \cdots + (n-1)^3) = \frac{(n-1)^2 n^2}{4n^4} = \frac{(n-1)^2}{4n^2}$$

(as illustrated below) and similarly

$$m_n = \left(\frac{1}{n} \times \left(\frac{1}{n}\right)^3\right) + \left(\frac{1}{n} \times \left(\frac{2}{n}\right)^3\right) + \cdots + \left(\frac{1}{n} \times 1^3\right)$$
$$= \frac{1}{n^4}(1^3 + 2^3 + \cdots + (n-1)^3 + n^3) = \frac{n^2(n+1)^2}{4n^4} = \frac{(n+1)^2}{4n^2}$$

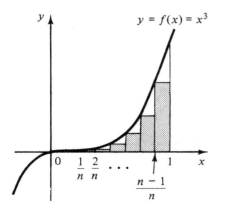

Since the sequences (l_n) and (m_n) both converge to $\frac{1}{4}$ it follows from the definition that f is integrable over $[0, 1]$ with integral equal to $\frac{1}{4}$.

2   (i) The required formula for $\sin r\alpha$ follows easily from the fact that

$$\cos(r\alpha - \tfrac{1}{2}\alpha) = \cos r\alpha \cos \tfrac{1}{2}\alpha + \sin r\alpha \sin \tfrac{1}{2}\alpha$$

and

$$\cos(r\alpha + \tfrac{1}{2}\alpha) = \cos r\alpha \cos \tfrac{1}{2}\alpha - \sin r\alpha \sin \tfrac{1}{2}\alpha$$

Hence

$$(\sin \alpha + \sin 2\alpha + \sin 3\alpha + \cdots + \sin n\alpha) \times 2 \sin \tfrac{1}{2}\alpha = (\cos \tfrac{1}{2}\alpha - \cos \tfrac{1}{2}\alpha) +$$
$$(\cos \tfrac{1}{2}\alpha - \cos 2\tfrac{1}{2}\alpha) +$$
$$\vdots$$
$$(\cos(n - \tfrac{1}{2})\alpha - \cos(n + \tfrac{1}{2})\alpha)$$
$$= \cos \tfrac{1}{2}\alpha - \cos(n + \tfrac{1}{2})\alpha$$

and the sum of the sines follows.

(ii) The partition giving n equal strips in $[0, \pi/2]$ is $\{0, \pi/2n, 2\pi/2n, \ldots, (n-1)\pi/2n, \pi/2\}$. Using the above formula for the sum of sines this leads to the following under- and over-estimates in L and M:

$$l_n = \frac{\pi}{2n}\left(\sin 0 + \sin \frac{\pi}{2n} + \sin \frac{2\pi}{2n} + \cdots + \sin \frac{(n-1)\pi}{2n}\right)$$

$$= \frac{\pi}{2n}\left(\frac{\cos \dfrac{\pi}{4n} - \cos \dfrac{(n-\frac{1}{2})\pi}{2n}}{2 \sin \dfrac{\pi}{4n}}\right)$$

$$= \frac{\dfrac{\pi}{4n}}{\sin\left(\dfrac{\pi}{4n}\right)} \times \left(\cos \frac{\pi}{4n} - \cos \frac{(n-\frac{1}{2})\pi}{2n}\right) \to 1 \quad \text{as} \quad n \to \infty$$

since cos is continuous and $x/\sin x \to 1$ as $x \to 0$: l_n is illustrated below.

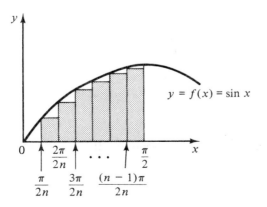

Similarly,

$$m_n = \frac{\pi}{2n}\left(\sin\frac{\pi}{n} + \sin\frac{2\pi}{n} + \cdots + \sin\frac{\pi}{2}\right)$$

$$= \frac{\pi}{2n}\left(\frac{\cos\dfrac{\pi}{4n} - \cos\dfrac{(n+\frac{1}{2})\pi}{2n}}{2\sin\dfrac{\pi}{4n}}\right)$$

$$= \frac{\dfrac{\pi}{4n}}{\sin\!\left(\dfrac{\pi}{4n}\right)} \times \left(\cos\frac{\pi}{4n} - \cos\frac{(n+\frac{1}{2})\pi}{2n}\right) \to 1 \quad \text{as} \quad n \to \infty$$

Hence both the sequences (l_n) and (m_n) converge to 1 and it follows that the sine function is integrable over $[0, \pi/2]$ with integral 1.

3　　　　For the partition $a = a_0 < a_1 < \cdots < a_n = b$ which divides $[a, b]$ into n equal pieces we have

$$m_n - l_n = \sum_{r=1}^{n} \underbrace{(a_r - a_{r-1})}_{= (b-a)/n}(\underbrace{\sup\{f(x): a_{r-1} \leqslant x \leqslant a_r\}}_{=f(a_r)} - \underbrace{\inf\{f(x): a_{r-1} \leqslant x \leqslant a_r\}}_{=f(a_{r-1})})$$

since f is increasing

and so

$$m_n - l_n = \frac{1}{n}(b-a)((f(a_1) - f(a)) +$$
$$(f(a_2) - f(a_1)) +$$
$$\vdots$$
$$(f(b) - f(a_{n-1})))$$
$$= \frac{1}{n}(b-a)(f(b) - f(a)) \to 0 \quad \text{as} \quad n \to \infty$$

To show that f is integrable over $[a, b]$ we will show that the sequences (l_n) and (m_n) both converge and have the same limit. By the theorem on page 198 we have (with the usual notation) $l \leqslant m$ for each $l \in L$ and $m \in M$ and (as we observed in the proof of that theorem) there exists some number α 'between' L and M. Therefore as $n \to \infty$ we have

$$0 \leqslant m_n - \alpha \leqslant m_n - l_n \quad \text{and} \quad 0 \leqslant \alpha - l_n \leqslant m_n - l_n$$

$$\begin{array}{ccc} \downarrow & \qquad \Big\downarrow & \downarrow \\ 0 & \therefore & 0 \\ & 0 & \end{array} \qquad \begin{array}{ccc} \downarrow & \Big\downarrow & \downarrow \\ 0 & \therefore & 0 \\ & 0 & \end{array}$$

and so both the sequences (l_n) and (m_n) converge to α. It follows that f is integrable as required.

In particular the function f given by $f(x) = [x^3]$ is increasing and so it is integrable over every interval $[a, b]$.

To find the integral over $[0, 1\frac{1}{2}]$ note that the partition

$$\{0, 1, \sqrt[3]{2}, \sqrt[3]{3}, 1\tfrac{1}{2}\}$$

gives under- and over-estimates of the integral which are actually equal.

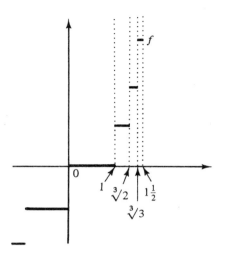

Hence the integral must be this common value and so

$$\int_0^{1\frac{1}{2}} [x^3]\,dx = 0 \times (1 - 0) + 1 \times (\sqrt[3]{2} - 1) + 2 \times (\sqrt[3]{3} - \sqrt[3]{2})$$

$$+ 3 \times (1\tfrac{1}{2} - \sqrt[3]{3}) = 3\tfrac{1}{2} - \sqrt[3]{2} - \sqrt[3]{3}$$

4   Following a similar procedure to that in the worked examples but with $[a, b]$ partitioned into n equal parts we obtain

$$l_n = \left(\frac{b - a}{n} \times a^2\right) + \left(\frac{b - a}{n} \times \left(a + \frac{b - a}{n}\right)^2\right)$$

$$+ \left(\frac{b - a}{n} \times \left(a + \frac{2(b - a)}{n}\right)^2\right)$$

$$+ \cdots + \left(\frac{b - a}{n} \times \left(a + \frac{(n - 1)(b - a)}{n}\right)^2\right)$$

which (with the use of our formula for the sum of squares) eventually simplifies to

$$l_n = (b - a)\left(\frac{a^2}{n} + \frac{n - 1}{n}\,ab + \frac{(n - 1)(2n - 1)}{6n^2}(b - a)^2\right)$$

$$\to \tfrac{1}{3}(b^3 - a^3) \quad \text{as} \quad n \to \infty$$

The over-estimates can be found similarly and they too tend to that limit. Hence the integral is $\tfrac{1}{3}(b^3 - a^3)$ as expected.

(In fact, as you may have begun to suspect, for a continuous function the sequence (l_n) of under-estimates obtained by considering the evenly-spaced partitions will always converge to the required integral.)

5   (i) and (ii) Note first that (as we saw in exercise 3(iv) on page 13) if A is a non-empty set which is bounded below then the set $B = \{-a : a \in A\}$ is bounded above and $\sup B = -\inf A$.

Now for the arbitrary partition of $[a, b]$ we have the under-estimate of g's integral given by

$$l = \sum_{r=1}^{n} (a_r - a_{r-1}) \times \inf\{g(x): a_{r-1} \leqslant x \leqslant a_r\}$$

$$= \sum_{r=1}^{n} (a_r - a_{r-1}) \times \inf\{-f(x): a_{r-1} \leqslant x \leqslant a_r\}$$

$$= \sum_{r=1}^{n} (a_r - a_{r-1}) \times -\sup\{f(x): a_{r-1} \leqslant x \leqslant a_r\}$$

$$= -\sum_{r=1}^{n} (a_r - a_{r-1}) \times \sup\{f(x): a_{r-1} \leqslant x \leqslant a_r\}$$

So for any $l \in L'$ we have $-l \in M$, the set of over-estimates of f's integral. The reverse implication follows easily and (ii) is proved in a similar way.

(iii) It follows from (i) and (ii) (and our comments about infs/sups) that

$$\inf L' = -\sup M \quad \text{and} \quad \sup M' = -\inf L$$

Now by the theorem on page 198 g is integrable if and only if

$$\inf L' = \sup M'$$
$$\| \qquad \|$$
$$-\sup M \quad -\inf L$$

But these are equal since f is integrable, and so g is integrable and

$$\int_a^b g = \inf L' = -\sup M = -\int_a^b f$$

In the next section we shall define 'lower' and 'upper' integrals and the result $\inf L' = \sup M$ will translate to

$$\underline{\int_a^b} (-f) = \underline{\int_a^b} g = -\overline{\int_a^b} f$$

Page 217

1        $F'(x) = (f'(x)g(x) + f(x)g'(x)) - f'(x)g(x) = f(x)g'(x)$
and so F is an indefinite integral of fg' and

$$\int_a^b fg' = F(b) - F(a) = [f(x)g(x)]_a^b - \int_a^b f'g$$

as required.

(i) $f(x) = x$, $g(x) = -\cos x$ gives $g'(x) = \sin x$ and

$$\int_0^{\pi/2} \underbrace{x}_{f} \underbrace{\sin x}_{g'} dx = \underbrace{[-x \cos x]_0^{\pi/2}}_{=0} + \int_0^{\pi/2} \cos x \, dx = [\sin x]_0^{\pi/2} = 1$$

(ii) $f(x) = \log x$, $g(x) = x$ easily gives the integral as 1;

(iii) $f(x) = e^x$, $g(x) = -\cos x$ gives

$$I = \int_0^{\pi/2} e^x \sin x \, dx = \underbrace{[-e^x \cos x]_0^{\pi/2}}_{=1} + \int_0^{\pi/2} e^x \cos x \, dx$$

and applying the rule with a fresh start of $f(x) = e^x$, $g(x) = \sin x$ gives

$$I = 1 + \left(\underbrace{[e^x \sin x]_0^{\pi/2}}_{e^{\pi/2}} - \int_0^{\pi/2} e^x \sin x \, dx \right) = 1 + e^{\pi/2} - I$$

and from this equation in I we get $I = \frac{1}{2}(1 + e^{\pi/2})$.

2   Let

$$F(t) = \int_{g(a)}^t f$$

so that $F' = f$ and

$$F(g(x)) = \int_{g(a)}^{g(x)} f$$

By the chain rule we have

$$(F \circ g)'(x) = (F' \circ g)(x) \times g'(x) = (f \circ g)(x) \times g'(x)$$

and so $F \circ g$ is an indefinite integral of $(f \circ g) \times g'$ and

$$\int_a^b (f \circ g) \times g' = (F \circ g)(b) - (F \circ g)(a) = \int_{g(a)}^{g(b)} f$$

as required.

(i) $u = g(x) = x^2$, $f(u) = \sin u$ gives the indefinite integrals

$$\int 2x \underbrace{\sin(x^2)}_{} \, dx = \int \underbrace{\sin u}_{} \, du = -\cos u = -\cos(x^2)$$
$$\quad\;\; \underbrace{}_{g'} \;\; \underbrace{}_{f \circ g} \qquad\qquad \underbrace{}_{f}$$

(ii) $u = g(x) = e^x$, $f(u) = 1/(1 + u^2)$ gives

$$\int \frac{e^x}{1 + e^{2x}} \, dx = \int \frac{1}{1 + u^2} \, du = \tan^{-1} u = \tan^{-1} e^x$$

(iii) $u = g(x) = \cos x$, $f(u) = 1/u$ gives the answer as $-\log(\cos x)$.

3   (i) For $0 \leqslant x < 1$

$$F(x) = \int_0^x \frac{1}{\sqrt{(1 - t)}} \, dt = [\sin^{-1} t]_0^x = \sin^{-1} x \to \frac{\pi}{2} \quad \text{as} \quad x \to 1$$

and so the integral is defined and has value $\pi/2$.

(ii) As $x \to \infty$ we have

$$F(x) = \int_0^x \frac{t}{(1 + t^2)^2} \, dt = \left[-\frac{1}{2} \frac{1}{1 + t^2} \right]_0^x = \frac{1}{2} \left(1 - \frac{1}{1 + x^2} \right) \to \frac{1}{2}$$

and so the integral is defined (or 'converges') and has value $\frac{1}{2}$.

(iii)   For $x > 0$

$$F(x) = \int_x^1 \frac{1}{t}\, dt = [\log t]_x^1 = -\log x \to -\infty \quad \text{as} \quad x \to 0$$

and so the integral is not defined (or it 'diverges').

4   As on page 47 x_n represents the shaded area shown below and so the sequence is increasing and bounded above by $f(1)$ (since the shaded region will fit inside the first rectangle).

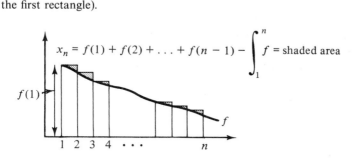

Hence (x_n) is increasing and bounded above and therefore convergent. If s_n is the nth partial sum of the series $f(1) + f(2) + f(3) + \cdots$ and

$$i_n = \int_1^n f$$

then we have that

$$i_n = s_{n-1} - x_n$$

and it's straightforward to check that (i_n) converges if and only if (s_{n-1}) converges. It follows that the improper integral is defined if and only if the series converges, as required.

(The result is easily adapted to

$$\int_k^\infty f \quad \text{and} \quad \sum_{n=k}^\infty f(n))$$

(i) With $f(x) = 1/\sqrt{x}$ we get

$$\int_1^n f = [2\sqrt{x}]_1^n = 2(\sqrt{n} - 1)$$

which does not converge: so the series diverges.

(ii) With $f(x) = 1/(x \log x)$ $(x \geq 2)$ we get

$$\int_2^n f = [\log(\log x)]_2^n = \log(\log n) - \log(\log 2)$$

which does not converge (so neither does the series).

(iii) With $f(x) = 1/(x \log x \log(\log x))$ $(x \geq 3)$ we get

$$\int_3^n f = [\log(\log(\log x))]_3^n = \log(\log(\log n)) - \log(\log(\log 3))$$

which does not converge.

(iv) With $f(x) = 1/(x(\log x)^2)$ $(x \geqslant 2)$ we get

$$\int_2^n f = \left[-\frac{1}{\log x} \right]_2^n = \frac{1}{\log 2} - \frac{1}{\log n} \to \frac{1}{\log 2}$$

and so the integral (and series) converge.

5   The case $n = 0$ says

$$f(x) = f(0) + \int_0^x f'$$

which is immediate from the fundamental theorem of calculus. So assume that $k > 0$ and that the result is known for $n = k - 1$. Then the last part of the expansion is as shown on the left below and integrating it by parts gives

$$\frac{1}{(k-1)!} \int_0^x (x-t)^{k-1} f^{(k)}(t) \, dt = \frac{1}{(k-1)!} \left[-\frac{(x-t)^k}{k} f^{(k)}(t) \right]_0^x$$

$$+ \frac{1}{(k-1)!} \int_0^x \frac{(x-t)^k}{k} f^{(k+1)}(t) \, dt$$

which simplifies to

$$\frac{x^k}{k!} f^{(k)}(0) + \frac{1}{k!} \int_0^x (x-t)^k f^{(k+1)}(t) \, dt$$

and the case $n = k$ is easily established.

In the case when $f(x) = \log(1 + x)$ we have

$$f^{(n)}(x) = \frac{(-1)^{n-1}(n-1)!}{(1+x)^n}$$

The size of the difference between $f(x)$ and its nth degree Taylor polynomial is therefore given by

$$d = \left| \frac{1}{n!} \int_0^x (x-t)^n \frac{(-1)^n n!}{(1+t)^{n+1}} \, dt \right| = \left| \int_0^x \frac{(x-t)^n}{(1+t)^{n+1}} \, dt \right|$$

Now integrals satisfy all the 'natural' inequalities. For example if $f(x) \leqslant g(x)$ for each $x \in [a, b]$ and f and g are integrable then

$$\int_a^b f \leqslant \int_a^b g$$

The interested reader can easily prove this and other simple properties. Hence for $0 \leqslant x \leqslant 1$

$$d = \int_0^x \underbrace{\frac{(x-t)^n}{(1+t)^{n+1}}}_{\text{denominator} \,\geqslant\, 1} \, dt \leqslant \int_0^x (x-t)^n \, dt = \frac{x^{n+1}}{n+1} \to 0 \quad \text{as} \quad n \to \infty$$

and for $-1 < x \leqslant 0$

$$d = \left| \int_0^x \frac{(x-t)^n}{(1+t)^{n+1}} \, dt \right| = \int_x^0 \frac{(t-x)^n}{(1+t)^{n+1}} \, dt$$

$$= \int_x^0 \left(\frac{t-x}{1+t}\right)^n \underbrace{\frac{1}{1+t}}_{\leqslant 1/(1+x)} dt \leqslant \frac{(-x)(-x)^n}{1+x} \to 0 \quad \text{as} \quad n \to \infty$$

$$\underbrace{\phantom{\left(\frac{t-x}{1+t}\right)^n}}_{\leqslant (-x)^n}$$

So in each case the difference tends to 0 as n increases and the Taylor series converges to $\log(1 + x)$ for each $x \in (-1, 1]$.

6   We saw in theorem on page 185 that the series

$$a_0 + a_1 x + a_2 x^2 + a_3 x^3 + \cdots \quad \text{and} \quad a_1 + 2a_2 x + 3a_3 x^2 + \cdots$$

have the same radius of convergence and that if the left-hand series sums to $f(x)$ for $x \in D$ then f is continuous on D and (apart from possibly at the endpoints of D) the derived series on the right converges to $f'(x)$ for $x \in D$. With a change of labels we can clearly take the right-hand series summing to $g(x)$ as our starting point and look at the 'integrated' series on the left. By the above comments it is clear that the two series have the same radius of convergence and that within the interval of convergence the left-hand series sums to a function which is an indefinite integral of g.

Applying this to the geometric series

$$\frac{1}{1+x^2} = 1 - x^2 + x^4 - x^6 + \cdots \quad x \in (-1, 1)$$

and 'integrating' we get

$$\tan^{-1} x = x - \frac{x^3}{3} + \frac{x^5}{5} - \frac{x^7}{7} + \cdots \quad x \in (-1, 1)$$

In fact this last series also converges for $x = 1$ (by the alternating series test on page 71 and so (again by the theorem on page 185) its sum is continuous throughout $(-1, 1]$. Therefore (keeping to $x \in (-1, 1]$)

$$1 - \tfrac{1}{3} + \tfrac{1}{5} - \tfrac{1}{7} + \cdots = \lim_{x \to 1} \left(x - \frac{x^3}{3} + \frac{x^5}{5} - \frac{x^7}{7} + \cdots \right)$$

$$= \lim_{x \to 1} (\tan^{-1} x) = \tan^{-1} = \frac{\pi}{4}$$

as required. This is *Gregory's formula* for π, named after the seventeenth century Scot James Gregory who did a lot of work on series and on π. We have now nearly finished, more-or-less as we started, with a series which will give us arbitrarily close approximations to π. Mind you this series converges pretty slowly: my computer gives the sum of the first 1000 terms as 3.140 592... and the sum of the first 10 000 as 3.141 492.... In fact the sum of the first 10^n terms seems to differ from π ($= 3.141\ 592\ldots$) by approximately 0.00...01, where the 1 is in the nth place; but that's another story altogether. And finally, to see that π is irrational:

7   We know that

$$0 \leqslant \frac{N^n \pi^{2n+1}}{n!} = \pi \frac{(N\pi^2)^n}{n!} \to 0 \quad \text{as } n \to \infty$$

and so, for some positive integer k, $N^k \pi^{2k+1}/k!$ is less than 1.

(i) By Leibniz' rule for differentiating a product

$$f(x) = \frac{N^k}{k!} x^k(x - \pi)^k$$

$$f'(x) = \frac{N^k}{k!} (x^k(x - \pi)^{k-1} + kx^{k-1}(x - \pi)^k)$$

$$f''(x) = \frac{N^k}{k!} (k(k - 1)x^k(x - \pi)^{k-2} + 2k^2 x^{k-1}(x - \pi)^{k-1}$$

$$+ k(k - 1)x^{k-2}(x - \pi)^k)$$

$$\vdots$$

all of whose terms are 0 at $x = 0$ and $x = \pi$

$$f^{(k)}(x) = \frac{N^k}{k!} (k!x^k + \underbrace{\quad \cdots \quad}_{\substack{\text{terms involving} \\ x^i(x - \pi)^j \text{ for} \\ \text{positive } i, j}} + k!(x - \pi)^k)$$

$$= \pm N^k \pi^k = \pm M^k \text{ at } x = 0 \text{ and } x = \pi$$

$$\vdots$$

$$f^{(2k)}(x) = \frac{(2k)! N^k}{k!} = \text{integer}$$

Hence f and all its derivatives have integer values at $x = 0$ and $x = \pi$.

(ii) For each $x \in [0, \pi]$

$$0 \leqslant f(x) \sin x = \frac{N^k}{k!} x^k(x - \pi)^k \sin x \leqslant \frac{N^k \pi^k \pi^k}{k!} = \underbrace{\frac{N^k \pi^{2k+1}}{k!}}_{<1} \frac{1}{\pi} < \frac{1}{\pi}$$

and so

$$0 \leqslant \int_0^\pi f(x) \sin x \, dx < (\pi - 0) \frac{1}{\pi} = 1$$

In addition the integral, I say, is clearly not 0 (for example the partition $(0, \pi/4, 3\pi/4, \pi)$ will certainly lead to to a positive under-estimate). Hence

$$0 < I < 1$$

as required.

(iii) By repeatedly integrating by parts

$$I = \int_0^\pi f(x) \sin x \, dx = \underbrace{[-f(x) \cos x]_0^\pi}_{\text{integer (by (i))}} + \int_0^\pi f'(x) \cos x \, dx$$

$$= \text{integer} + \underbrace{[f'(x) \sin x]_0^\pi}_{\text{integer (0)}} - \int_0^\pi f''(x) \sin x \, dx$$

$$\vdots$$

$$= \text{integer} \pm \int_0^\pi \underbrace{f^{(2k)}(x)}_{\substack{\text{constant} \\ \text{integer}}} \sin x \, dx$$

$$= \text{integer} + \text{integer} \times [\cos x]_0^\pi = \text{integer}$$

Therefore our assumption that π was rational has led to the obvious contradiction

$$0 < I < 1 \quad \text{and} \quad I \text{ is an integer}$$

Hence π must be irrational, which is an entirely rational place at which to stop.

Index